GW00675460

Contents

Why me?

Why me?

SCIENCE AND SPIRITUALITY AS INEVITABLE BED PARTNERS

DR. PATRICK QUANTEN
& ERIK BUALDA

Copyright 2017, Patrick Quanten and Erik Bualda

All rights reserved. No part of this publication may be reproduced, stored in a retrieval system, or transmitted in any form or by any means, electronic, mechanical, photocopy, recording or otherwise, without prior permission from the authors.

Illustrations by Erik Bualda
Cover design by Erik Bualda and Tanja Prokop
(www.bookcoverworld.com)
Typeset in Arno Pro by Raphaël Freeman, Renana Typesetting
Book printing advice by Joe Gregory (www.rethinkpress.com)

ISBN: 978-908-278-541-8

Why We Wrote the Book and Why You Should Read It

In October 2000 I began to work on the pre-designs for our home. The intention was to build a modern house where the orientation of the building, sun movement and total functionality would be taken into account. I showed the concept to William Gijsen, a psychotherapist and author of several spiritual and philosophical books, to ask his opinion. When I told him I wanted a glass dome shaped like a pyramid he encouraged me to investigate the pyramid form in order to use my findings directly within the building. His remarks reverberated around my head for several days until I decided to study "the pyramid". The first question was, of course: "Where do I begin?"

The only thing I knew about pyramids was that they could be found all over the world, and that the best known is the pyramid of Cheops, also known as the Great Pyramid, situated in Egypt. And that is where I started!

I found an enormous amount of information about this particular pyramid in libraries and on the internet on its measurements, orientation, fractions, various cuts and connecting lines, and much more. An awful lot of people had already investigated the Great Pyramid looking for answers to the questions "why" and "how", but after reading their findings and absorbing the information provided, I was left with an unsatisfied feeling. I knew I needed to do my own research. And just as had happened to so many before me, the pyramid bug had bitten me too.

More than fifteen years later, 1,000 maps and around 33,000 documents later, I am still infected, but sharing my thoughts and insights with Patrick has resulted in a different understanding of the information held within the pyramid.

Erik

For me, the journey began with a strange telephone call from someone telling me he was doing research into pyramids and the light spectrum, who then asked if he could come and see me. That was Erik. All my life I have aimed to understand factors associated with health, and have worked on explaining disease patterns. At first, I couldn't comprehend what Erik wanted and what he expected from me as I knew nothing about pyramids. But my curiosity was aroused.

Through our sporadic contact over the years we developed lines of parallel thinking that would not have been possible without that strange phone call. This engendered an interaction between modern, advanced and expansive thinking whereby, via scriptures of various other sources, we came to know the traditional values of ancient cultures.

Inspired by Erik's many questions, I became open to looking at the focus of my studies over the years in a broad and detached way, especially at the energetic and physical structure of the human as well as the functioning of that structure. Erik wanted all the i's dotted and the t's crossed with regard to this area of investigation, which was so refreshing compared with the medical world I knew. Somehow I got the feeling that he was right and that it was important to line everything up and find the natural connections that make up evolution and everything within it. The search had to be continued.

My own insights about illness and health, and specifically about birth defects and children's diseases, would not have emerged without the stimulation offered by the work Erik was doing.

Patrick

Why would someone be interested in reading this book? Let's be honest: it's a work that claims to provide answers to the basic questions of life, written by two unknowns. Who in their right mind could take this work seriously? Surely it must either be a hoax or the work of a couple of megalomaniacs.

On the one hand, we have an interior and industrial designer, Erik, who may be good at drawing pictures but has no background or schooling in Egyptian mythology or the Great Pyramid. Yet, here he is telling us all about it, and about how we should interpret the specific features of the building and the stories surrounding it. Who the hell does he think he is?

The other one is no better. Patrick gave up being a medical doctor to pursue crazy theories about health and inflicting his hands on suffering bodies and his voice on tired brains. He tells us what the Bible means, how Ayurveda relates to modern science, whilst he obviously is not an expert on either. Two ordinary guys who have led double lives, so to speak, in search of whatever they regard as the truth. Now they emerge from their dark, hidden, insignificant lives to enlighten us on questions such as:

- Where does life come from?
- What is the purpose of life?

- What is the role of the human being within the cosmos?
- How does matter emerge from energy?
- What are the direct links between the human energetic field, human chakras and human tissues?
- What are the basic laws of life: the rules upon which the entire universe is built and on which it functions?

It sounds to me as if we have the pretence to state that we have found important missing links, which could, in fact, prove to be needed in order to formulate the unified theory, the ultimate goal of Albert Einstein and every serious physicist that came after him. And all this done by two people without the proper qualifications or training? Impossible.

Well, let's not forget that most major breakthroughs in man's understanding of life have come as a result of input from people coming from outside the scientific community that had been looking to solve the puzzle. So, in the first place it isn't unheard of. It really happens. But why would we specifically believe that this could be one of those moments in history?

If the information in this book is to be of value, it needs to comply with everything that the contemporary scientific world already knows. What does that mean in real terms? It means that the information should fit neatly within the framework of our understanding of the universe as a whole. That understanding is based on the following principles:

- Everything is energy: all matter is created from energy, and is, in fact, nothing but condensed energy.
- Energetic information will be manifest within the kind of matter that is being produced and will determine the way the matter functions.
- All energetic influences play a part in creation; all energetic influences are ultimately construed from cosmic energies.
- The universe, the creation human beings belong to, is made

according to a simple template that keeps being repeated (chaos theory).

- Everything within the universe is cyclical; everything has already happened before and will happen again.

From my point of view, having been educated within the medical profession, this framework of life doesn't make any sense, because doctors are made to believe that everything has a material cause. They, as the rest of us, are convinced, because they have been told repeatedly, that there is a mechanical way the human being, the human body, functions. Scientists have told us for nearly a century that everything is made up of energy and that all explanations as to questions about life must be found in understanding energy, not matter. But believing this means we are misguided and believe the opposite of what is in fact true!

If our media and so-called "popular science" are not telling us the truth, how can we be expected to understand our own lives? How can we ever hope to live a healthier life if all our efforts to counteract diseases are based on the *Interaction of Matter*, whilst matter only acts out energetic resonances? In other words, we are seriously being lied to. No one in any position to do so shows the influence the mind has on the functioning of the bodily machines we operate. Nobody tells us that science proved thirty years ago that your emotions determine which proteins are produced in your body and that altering your emotional status will change your physical tissue (Candace Pert), allowing for diseases to "disappear" overnight. In addition, New Biology has proven that every cell of the body receives the same energetic information from our environment and translates it into an appropriate response, coordinating the whole human system as a physical unit. The communication system within the matter of the body is energy (Bruce Lipton). Albert Einstein himself proved that when you test something you are an integral part of the result. In other words, the way you observe, the way you organise the test and the expectations you have regarding the test results are all part of the material final result. So why is the medical profession still relying on tests to tell you what is wrong with you?

The truth has something to do with the concept that *We Are Made from Energy and We Function Energetically*. This means that if a human body has been created with obvious faults, there is an energetic explanation for it, and if a human body malfunctions and becomes ill, there is an energetic cause for it.

Wouldn't we want to know how that works? Wouldn't you, as an individual, want to know what really causes you to malfunction? Experience with using matter as a working material to provide health benefits to people shows us that this approach doesn't work at all. No matter the recommended treatment, Western medicine and its alternatives teach us that *All Therapies Work, but Not for Everybody and Not All the Time*. This means that none of the treatments correctly identify the cause of disease. We can now understand that if we continue to look for such cause in the material part of creation, we are not going to find the energetic reasons for the matter to function in the way it does. In short, we are looking in the wrong place, despite scientists having told us of this fact a hundred years ago!

And what's more, if authorities are not willing to tell you the truth, you will need to find out for yourself. That's how you end up with guys like us doing the work we do. That's all we started out doing: looking for answers to simple questions we felt we were not receiving. From that point onwards a journey began that no one could predict. You don't start out thinking: "I'll show them. I am going to be the greatest scientist that ever lived." No, you look for your own answers to things you want to know, and in order for you to know them, you have to find out about them for yourself. You'll find bits of interesting information everywhere. There are lots of people around the world doing the same thing, piecing together bits of the universal jigsaw puzzle. And the more you look, you can see there are an increasing number of weird things that you begin to notice. Bit by bit you'll be drawn into a world that is totally alien to you. And yet, in order to find answers for yourself, you will need to familiarise yourself with this world. But as you came in freelance, you are not given the guided tour all the regular students are. This means you take twists and turns that the experts-to-be are not allowed to take, and you end up in places no level-headed thinker would. On the way, you gather bits of information that may mean very little in the place you have found them, but they link up with information from completely different sources that could well provide the missing insight. A directed and well-guided education steers you in a predetermined direction and delivers almost predictable results. It is like a guided visit to China, which leaves you with the impression that all is hunky-dory under the Eastern sun.

Now, just suppose that what you have found is clearly more of the truth than the truth you have lived with so far. Wouldn't you want to tell others about it, for no other reason than that there may be other people out there searching for something similar? But how do we know that this book contains more of the truth than the everyday truth we are meant to live with?

I have already mentioned that the first principles relate to the already existing knowledge of the universe. Well, it clearly fits with that. But what else has it got?

We have grown up with the belief that ancient stories are just that – stories. We are told that we do not need to take them literally because at the time the stories were written, people's minds were not very well developed and they didn't know the things we know now. The stories, we are told, were the ways primitive people tried to explain the world around them in a child-like way. You know what becomes really disturbing when you journey in search of your own truth? You soon find out that not only did these primitive people actually create structures that we, the wise ones, are incapable of reproducing, such as the Great Pyramid, but also that our own modern scientists are closing in on proof of the authenticity of the literal meaning of the stories. But nobody tells you this! And if you do tell someone, they will try to commit you to a mental institution.

Through self-study, personal research and a lot of personal experience (including personal experimentation) we were gradually able to connect information from a wide variety of sources. This then started to point in one and the same direction. When you become aware of the one-pointed direction of information, you know that you are on the way to finding at least some of the answers. It is then that you start to get excited. And it is then that you have to become quiet, because your chances of becoming a human renegade shoot through the roof. You are officially going insane!

Ancient information collected from all over the world, be it from documents, recorded stories and significant buildings, can be linked to information derived from modern science, including within the fields of mathematics, physics, chemistry and biology. When such information from different sources fits together effortlessly, it can provide insight that does more than just provide answers to questions you seek. It also shows you that your interpretations and deductions have significance and meaning. Further, when we link old with new information, and equally and effortlessly include information from the west and the east, it truly feels as if we have gone global with our knowledge. Spirituality connects with physical matter, as information that spans place and time joins together. These connections, and our understanding of them, are essential for the whole of humanity to raise its consciousness and evolution platform, and to be ready for a new level – a new era.

The information in this book aims to take humanity into the new era of 2012 – a new level of human development. It is the culmination of knowledge and insight at the end of one cycle, leading to the beginning of another. We hope you are able to enjoy it as we do, witnessing the changeover with eyes wide open and minds bubbling with excitement.

There is great potential for the impact of the information within the book to be felt by every individual. First of all it proves that we are all made differently, which means the claim that *One Solution Fits All* is ludicrous, and therefore disavows the Western concept that a particular treatment will cure everyone from a certain disease, or that a particular lifestyle will be healthy for everyone, or that a particular social or support structure will serve everyone. This is clearly not the case, and now we know why. Hence, there is no point in further pursuing the idea that community solutions can be found to individual problems.

Secondly, the book shows that the way the individual functions is an interaction between the way the individual has been created (the inner world) and the outer influences coming from the environment; those being the individual's surroundings (the outer world). Only by making changes to either of these can we change the way the individual functions. This means that the cure for all diseases requires the individual to change the balance of the interaction between their inner and outer energies.

Thirdly, because we know the basic frequencies that create the structure of the human being, we can know the basic harmony upon which an individual has been created. This means we can help the individual to regain harmony by encouraging the changes he/she needs to make through exposure of their system to very specific stimulating frequencies in the places that use this information to create the physical parts of the human being.

Not only will human health never be as far removed from us as it has recently become, but we will gain an understanding of life that makes health a simple individual choice. Our definitions of health and disease will change forever. We will gain more power over our own lives by understanding the basic influences that create life itself and make it function. This, in turn, will result in a change in the way we structure society: it will move from a group-centred society to an individual-centred society, but with the added bonus that individuals will no longer need to be selfish and greedy. They will understand that all they need is within their grasp and that more will not serve them at all.

So you see, with this information we are truly ready for the new era, as it will change the structure and function of the individual as well as the group structure we live in. Let all of them read the book, for their sanity and ours!

The old Egyptians called the Great Pyramid "Ta Khut", meaning "the light". There must be a reason why they called Gizeh Rostau the gate to the other world. In this book we investigate how light came into being and what it has meant in terms of evolution, from the Big Bang to our present society. At various stages we will refer to our study and findings in relation to the Great Pyramid. This will then be dealt with in detail in the second book, Part II, where we reveal how our search brought us to "the gate to the other world".

Chapter 1
The Creation

I find it very interesting that every culture has a story about "how it all began". People have always felt the need to comprehend the creation of the world and how we came to be. In order to gain such an understanding, it is necessary to know where Today came from. In other words, what is behind it? What is its history? Once you understand how yesterday became today you probably have a good idea what tomorrow could look like. Understanding the creative process is the key. Ultimately, that is exactly what science has put forward as its goal. That is why Einstein wasn't happy with his theory of general and special relativity, because he knew that it was wrong to have one theory for the small bits in life (quantum physics for the microscopic world) and another for the large components of life (cosmic theory). He kept working to find the unifying theory, to understand all phenomena of the universe, large and small, based on the same model. And since then, scientists across the world still concern themselves with unifying various theories.

It is only when you begin to comprehend the real meaning of things that you can see order in what first appears as chaos. It is only when you truly understand that you are able to view the roots of your old belief system, which appeared until that moment as the logic and truth itself. To find the real inner truth will always shake society because society has built itself upon a structure based on a perceived truth, which in fact is only an illusion of the truth. As long as the structure is not built on the real truth, the true reality of life, the true laws of nature and the universe, it will eventually clash with the truth that lies inside. This will always lead to a crumbling of the structure. This is why all major cultures we have known have disintegrated, no matter how sophisticated they were.

If we truly want to understand, it is imperative that we join together all knowledge humans have already gathered. A scientific approach that holds onto dogma and refuses to look at certain observations and knowledge is deemed to remain steeped in ignorance. Not examining what cannot be explained or appears dangerous to the structure is going to keep the structure standing, at least for now, but it slows down the intellectual and spiritual development of the population. Authorities hold on to their power by keeping people ignorant. Creating "experts" helps to perpetuate the *Illusion* of knowledge and ensures a high regard for persons that people *Believe* have knowledge. However, real knowledge is not restricted by education and cannot be measured by human standards. Don't forget that Jesus was but a carpenter and still his knowledge far outreached that of the scholars of his time and place.

We started this journey with the intention to integrate knowledge from different corners of science. Don't be deterred by the origin of specific wisdom and be careful not to dismiss anything before you have had the chance to study the detail. Allow information to meld together and start an exchange.

Egyptian Mythology

The different creation myths in Egyptian mythology have some elements in common. They all hold that the world has arisen out of the lifeless waters of chaos, called *Nu*. They also include a pyramid-shaped mound, called the *Benben*, which was the first thing to emerge from the waters. These elements were likely inspired by the flooding of the River Nile each year. The receding floodwaters left fertile soil in their wake, and the Egyptians may have equated this with the emergence of life from primeval chaos, with the imagery of the pyramidal mound having been derived from the highest mounds on earth emerging as the river receded.

The sun was also closely associated with creation, and it was said to have first risen from the mound as the general sun god *Ra* (also called *Re*) or as the god *Kephri*, who represented the newly-risen sun. There were many versions of the sun's emergence, and it was said to have emerged directly from the mound or from a lotus flower that grew from the mound, in the form of a heron, falcon, scarab beetle or human child.

The story of Heliopolis comes to us on a papyrus amidst bad omens. It was handed as a gift to Rhind in 1861 by the then British Consul in Luxor. Mustafa Aghs obtained the document from the exhibition of royal mummies at Deir el-Bahari.

In Heliopolis, the creation was attributed to *Kephri*, a deity closely associated with Ra, who was said to have existed in the waters of Nu as an inert potential being. The Heliopolitan myth described the process by which he "evolved" from a single being into this multiplicity of elements. The process began when Kephri appeared on the mound and gave rise to the air god *Shu* and his sister *Tefnut*, whose existence represented the emergence of an empty space amid the waters. To explain how Atum (Kephri) did this, the myth uses the metaphor of masturbation, with the hand he used in this act representing the female principle inherent within him. He is also said to have "sneezed" and "spat" to produce Shu and Tefnut, a metaphor that arose from puns on their names. Next, Shu and Tefnut coupled to produce the earth god *Geb* and the sky goddess *Nut*, who defined the limits of the world. Geb and Nut in turn gave rise to four children, who represented the forces of life: *Osiris*, god of fertility and regeneration; *Isis*, goddess of motherhood; *Set*, the god of male sexuality; and *Nephthys*, the female complement of Set. The myth thus represented the process by which life was made possible. These nine gods were grouped together theologically as the Ennead, but the eight lesser gods, and all other things in the world, were ultimately seen as extensions of Kephri.

Even before they were born, Osiris and Isis were man and wife, and they had a son named *Horus*. Set and Nephthys also had a son, called *Anubis*. Kephri gave Osiris a wonderful gift as he was made from the same substance as his grandfather, and thus became the incarnation of the Creator. Later, he would rise from the dead and become the saviour of humankind.

And finally, humanity was created from the tears Kephri shed, and humans were thus the direct descendants of the Creator, and not a product of earth.

Most of the Egyptian gods were lookalikes and could only be distinguished by the use of symbols or their appearance. Symbols could be things they wore, such as the crown worn by Osiris or the falcon head of Geb, or they could be things they held in their hands, symbolising their power and status.

Other ways of recognising the gods was through their association with animals. A god would be depicted as an animal or as a hybrid; a combination of human and animal form. For instance, the god Toth has the body of a human but the face of an ibis. These were well-chosen representations. The goddess Taweret is represented as a hippopotamus; a dangerous animal protective of her children. Taweret was also the protector of pregnant women.

Ra, or Re (depending on the various sources), was the sun god in Egyptian mythology; an important god and one of the most worshipped.

Ra ruled the heavens. When he was too old to rule the earth he took to the skies, where he travelled in his boat alongside the god Ma'at and others.

Ra was also the ruler of earth. A myth tells the story of Ra having once been a pharaoh, long before the pharaoh dynasty.

Ra also appeared in the underworld. As he travelled the heavens in his boat, so did he travel the underworld, where he was equal to Osiris.

Ra was also known as the Creator. Many stories about creation have circulated throughout Egyptian history, such as the one from Heliopolis in which Ra created the earth.

Ra has been honoured throughout Egypt's history. He has been depicted in many different forms, such as a sun alongside a cobra with or without stretched-out wings. Often we see him in a human body with a pig's or ram's head. He can, however, appear as a variety of animals, such as a ram, phoenix, snake, bull, cat, lion or compilations of animals.

Osiris is the son of Geb, the god of the earth, and his sister Nut, the goddess of the heavens. Osiris became the first king of Ancient Egypt. His wife, who was also his sister, was Isis. Osiris' attributes included a shepherd's crook and a flagellum. He is said to have reigned as a good shepherd, until he was killed by his brother Set, who wanted to become master of the legacy. Set cut the body of Osiris into *14 Pieces* but his wife Isis found them and eventually managed to ensure eternal life for Osiris. Osiris' son Horus took revenge for his father's death on Set. Plenty of myths can be found around this theme.

The myth of Osiris is probably the best-known story of Ancient Egypt and tells the tale of his death and resurrection, symbolising the cycle of life and death, and of sunrise and sunset. Osiris was not only god of all kings offering the power of life to all pharaohs, he was also the personification of the fertility of the land and the mind of the cycle of vegetation. As Judge of the Dead he granted a new life to those who deserved eternal life by the pure way they had lived.

Ancient Science

In the *Rig Veda*, the oldest writings of India (around 7000 years old) it states that there are three fundamental forces.

First, there is the principle of energy (The Father), which gives power, speed, direction and motivation. In modern science, this is seen in the fact that everything is energy and that all matter that appears to us as "solid" is in fact an interplay of various forces.

The second force is the principle of light (Holy Ghost), or radiation. Energy is light, and when it moves, it changes and emanates light and heat. There is a natural light attached to all energy; all energy, including material stuff, emanates energy/light.

The third force is the principle of cohesion (The Son). In all manifestations there is unity. There is a coming together of forces in a singular rhythm of harmony; as a "substance" solidifies the various energies of all the parts blend together in one harmonious "movement". Love is the force that unites everything.

In the Vedas, the spirit of life is symbolised by the god Indra, the dragon slayer and the mighty hand behind the lightning bolt. The spirit of light was Agni, the god of fire, the divinity of vision and of sacrifice. The spirit of love was worshipped in Soma, the nectar of immortality.

These three forces were also symbolised by the three elements of air, fire and water. According to ancient mythology, in the beginning heaven and earth were one. There was no space between them in which living beings could manifest. Then, by the will of the Creator, the gods came into being and separated heaven and earth, drawing apart the two firmaments. In the space between, they set in motion the life force to allow creatures to come into being. This life force became the atmosphere in which the elements of air, fire and water, as wind, sun and rain, provided for the development of life.

In effect, the teaching talks about five elements, not three. Created by the three forces, the elements are air, fire and water, as well as ether and earth. Following the energy lines down towards the creation of matter, we find the three forces represented by the three basic energies that comprise the mind: Raja (energy), Tamas (cohesion) and Ojas (light). From these the five elements appear. Energy creates FIRE. Cohesion creates EARTH. Light creates ETHER.

In between ether and fire, AIR appears and in between fire and earth the element WATER is created.

Furthermore, the Vedas teach us that there are seven types of tissues, which arise during the materialisation process, and they are placed in a specific order. The plant, like the human being, and the universe itself, is composed of these seven types of "tissues". The juice of the plant corresponds to the plasma tissue of the human. The resin of the plant is its blood. The softwood equals the human muscle as a tissue. The gum of the plant is the fat of the human being. The bark is the bones. The leaves are its marrow and nervous tissue. The reproductive tissue of the plant is what we find in the flowers and the fruits.

The semen or reproductive tissue is the essence of all bodily tissues and contains the power of reproduction within itself as well as the power of rejuvenation.

Not only is the formation of the seven tissues the same for all living creatures, but within each level, as it is for plants and animals and humans, the development of the tissues themselves follows the same plan of growth, right from the creation of light itself. The flowering tree shows these tissues in their most developed state. The tree is to the plant world what the human being is to the animal kingdom.

The tissues form from the least dense to the most dense, whereby a next dense level is formed from the previous layer. The first tissue, juice or plasma is the first "new" tissue formed out of the reproductive tissue, the seed. Through densification out of juice the next tissue arises, which is the resin for the plants and the blood for the animals. From this one the softwood or muscle tissue arises. Each layer of tissue develops through further "hardening" of the existing tissue, until the last tissue is formed, which is the seed, and from which a new cycle can begin. These "tissues" are not the organ tissues we encounter inside the body; these are names that refer to the characteristics of the seven energy levels that matter consists of. We shall refer to the beginning of materialisation as the *Primary Tissues*.

Creation starts with the interaction of the three essential forces within the universe. These forces become manifest after the separation of heaven and earth, or in other words, once the stillness and peace, the unity of heaven has been broken. From the source, three forces arise, and from these all other things are created. All matter arises from densification of energy in a well-defined order.

The Bible

The biblical texts were not written by scientists for scientists; they were written by people from ancient times. The Genesis story, written by Moses around 1450 BC, was put to paper during the Bronze Age. Primitive stones and bronze tools were used to work the land, to make weapons and to build shelters. Writing, as we know it, had only been introduced a short while before, but was not yet a well-known activity and skill. The emphasis was still on finding food to survive. Natural science was limited to personal observation. Most people in those days believed in a multitude of gods. They also worshipped natural phenomena such as the sun, moon, stars, fire and water.

Moses, the author of the story of The Beginning in Genesis, had been educated in Egypt. The old Egypt flourished from 3000 BC to the first centuries AD. The Egyptians' success arose from the irrigation of the Nile valley, the early development of writing, trading with neighbouring areas and their military power. The Egyptians believed in a complicated network of gods and life after death. The Egyptian background of Moses can also be traced to his choice of words in the Hebrew texts.

In Genesis there is an emphasis on the "days" of creation (translated from the Hebrew "*Yom*"). In modern times we mostly talk of these days as containing 24 hours. It is not clever, nor important, to justify the existence of God and prove the truth in the Bible based on these "days". Bearing in mind the background of the Israelis (including Moses), it is easy to see that events from a past long gone are difficult to put into a realistic timeframe. The use of "*Yom*" to describe the process of creation seems a logical choice in order to bring the difficult concept into the story and the language so that people can make sense of it.

Genesis tells the story of how God created the earth and all the life it contains in six yoms. The old Hebrew had a limited choice of words (about 6000), and Moses used the word *Yom*, which is usually translated as *Day*, but in the Bible it has a variety of meanings. Specifically, it appears 2274 times and is translated as "days" 2008 times, 64 times as "time", 37 times as "chronicle", 32 times as "daily", 17 times as "forever", 18 times as "year", 10 times as "continual", 10 times as "when", 10 times as "if", eight times as "whilst", eight times as "full", four times as "complete", and 44 times in various other ways. Whether or not yom is a day of 24 hours or any other description of a time period, the first chapter of Genesis shows remarkable insights, which are totally unknown to Moses and his contemporaries.

There is one God: Most ancient religions, including Egyptian, believed in a multitude of gods. Genesis defines the concept of the one God, although the name used (Elohim) in Genesis 1 is a plural, which is used in combination with the singular verbal form.

God is outside his creation: The gods of ancient times were represented by natural phenomena, such as the sun (Egyptian god Ra) or the moon (Egyptian god Toth), within the creation. According to Genesis, God created the universe: "*In the Beginning God Created the Heavens and the Earth*". (Genesis 1:1)

There was a beginning: Most ancient religions believed that the universe had always existed. Even in the middle of the 20th century scientists still believed that the universe had always been there. During our time, science, especially astronomy, has found proof that the universe has a known beginning (Big Bang Theory). How could Moses have known this?

It didn't happen by accident: It happened because God made it happen (*"God Created..."*). Even the Big Bang Theory cannot explain *Why* it happened, only that it did.

The primary conditions were not right for life: In the beginning, whatever the exact atmospheric conditions might have been, earth was unable to sustain life, and scientists agree. Genesis confirms that after the creation of the earth several progressive steps were needed before the right conditions would occur.

Non-life pre-empted life; vegetation came before animal life: According to the book Genesis, God created plant life at the end of the third day. This corresponds with the conclusions of the natural sciences that amino acids are required to form proteins or RNA, which are the physical building blocks of living organisms. Vegetation is necessary to create oxygen in the atmosphere, which is a requirement for animal life to appear.

Simple preceded complex: From fossils we have been able to conclude that simple organisms lived long before the more complex life forms appeared. Genesis describes the same sequence of events.

Human beings appear last: According to scientific studies, human beings appeared not so long ago in geological terms; they came after all other known life forms. Genesis tells us that humans were created on the last day.

The story of creation begins with the creation of light. Everything else therefore manifests from light. The universe we live in is a light universe, or, to be more precise, a creation out of an electromagnetic force of which "light" is a creative part. If we want to understand the creation of the universe, and the life that develops within it, we need to get an insight into the effective role that electromagnetism plays in the creation of matter. This is something we will look at in more detail later.

On the first day, God created heaven and earth. This was essentially the ripping apart of the original unity of energies. It was in this in-between space that light began, and it was in this same space that life could now develop. The earth (as a representation of the earliest step in the creation of the universe) had been given a "skin", which separated it from the rest of the universe. This was the very first tissue, the first layer, of the universe. Matter, in its earliest from had been separated from non-matter.

On the second day, God separated the waters from the land. Here the oceans, the seas and the rivers were created. These formed the blood vessels, the bearers of nourishment, of the universe. This was the second tissue, called "blood".

On the third day, God created plant life in the sea and on land. This was the third tissue of the universe. This was the first tissue to have a denser form, but movement was still an important aspect of the tissue itself; the tissue was fixed into position but flexible. In Darwin's theory, we are told that fish and reptiles sprang from plankton, which is an ultimate form of plant life. Although it belongs to the previous layer of development, plankton shows some animal properties, and it was clearly a platform to the next level of creation. The third tissue layer is called "muscular tissue".

On the fourth day, God created fish, amphibians and reptiles. This is a reference to the lower animals, corresponding to the fourth tissue. All these animals procreate by laying eggs (it also includes birds). Animals form a denser tissue than plants, which is also manifest in the fact that animal tissue, call it meat, stores and concentrates elements. This facility allows for a higher nutritional factor, but it also means that toxins occur in higher concentrations in animal meat than in plants. In animals, energy gets fixed to a larger degree than it does in plants. This serves as a nutritional reserve for all tissues and protects against energy loss from the cold. This is the fourth tissue layer, called "fat".

On the fifth day, it was time for the great variety of animals. From the evolutionary theory we know that many species are a continuation of already existing species. So it is said that some reptiles have crawled out of the water and adapted to life on dry land. From the New Biology, however, it appears that this is not the case. New species emerge because others have finalised their selective evolution and the next step is a change in the morphogenetic field, which means that a new species is born suitable to its environment, the formation of which has been stimulated by its different expression of the seed information. This shows us that as more and more tissues appear, creation becomes denser and increasingly more complex in structure and simpler in energy. The tissues and the corresponding organisms become more highly developed. The less flexible fluidly the tissue becomes, the more complex the design of the organism and the more solid it manifests itself. Adaptability is traded in for stability! Kinetic energy is transformed into matter energy. The fifth tissue layer is called "bones".

On the sixth day, God created mankind. The sixth tissue is the most complex and developed tissue we have in the creation of this universe to-date. What makes man different from the previous layer of development, the animals, is our consciousness and ability to reason, which is a particular developmental step in the nervous tissue. There are animals that have reached the limit of this development stage in the previous layer, such as the chimpanzee and dolphin. This does not mean that we genetically come from these animals, but it means that we are a continuation of the development process. These animals form the top layer of the previous stage and we are the bottom layer of the next stage; they are, so to speak, the last note of one octave and we are the first note of the next octave. The sixth tissue layer is called "nervous tissue".

On the seventh day, God rested. This is because in the traditional Vedas the seventh tissue is the "reproductive tissue", which is the fruit of all created tissues that went before; all coming together, ready to pass the experiences and gathered knowledge on to the creation of the next seven stages. God rests because He now has to wait until His creation bears fruit before He can start a new cycle – a new world. Since the sixth tissue, mankind, has not matured fully, this creation is not yet bearing fruit, which is needed to seed the next universe, to move this universe on to a new one.

New Science

The universe is built out of basic particles, which are trapped by natural forces, of which gravity and the electromagnetic force are the most familiar. These act over long distances. Gravity keeps the large cosmic bodies in their orbits. The swirling electric currents within the earth's molten core give rise to magnetic fields that can swing a compass needle.

When you watch the effect of a magnet on iron filings you may wonder what agent communicates between them. We can give it a name, "the electromagnetic field", but this is not really an explanation. We are just inventing a label for the observed action over remote distances. Dirac discovered, amongst other things, that the electromagnetic field itself is ruled by quantum theory. Photons are particle-like bundles of electromagnetic radiation and they transmit an electromagnetic force as they flit between one charged particle and another. An electron oscillating back and forth in a radio antenna in London, for example, can cause a similar response within your radio at home. The communication medium between them is electromagnetic waves; radio waves that are also the movement of photons. Motion in one location gives rise to motion elsewhere as a direct result of electromagnetic waves.

In modern quantum field theory, all the forces, not just the electromagnetic force, are transmitted by bosons. There are two other forces that are less familiar because they are principally in and around the atomic nucleus. They are known as the "strong force" and the "weak force"; their names summarising their apparent strengths relative to the familiar electromagnetic force.

The strong force constructs protons and neutrons out of more basic pieces known as "quarks", which is the force that binds things together (such as gravity). While it is the weak force that makes the sun shine. It manages this by transforming its constituents into more stable combinations within the existing mass, whereby the mass releases this change in electrical charge into its environment. This is the radiation that is associated with the weak force.

The interplay between the weak and strong forces creates matter. The strong force basically holds the atomic nucleus together and, as its energy is squashed together into the atomic structure, the internal forces allow for a transformation of that matter, which is constantly being created and recreated. It is the weak force that guides this transformation and is responsible for the radiation that results from the change in electrical charges brought about by the transformation process.

Movement, on the other hand, is created by an interplay of electromagnetic forces and the force of gravity. Once again, one draws the energies inside (gravity) and the other emits energy (electromagnetism). These forces have an infinite playing field, unlike the limited force fields of the weak and strong forces. The latter ones act locally while the other two are extending their influence almost to infinity. Both are opposing forces and although they interact, they are not exclusive to time and place. The forces that make the matter are still penetrated by the forces that move the matter.

Watching a Mexican wave going around a sports stadium, for example, looks, from a distance, as if the surface of the spectators' area raises and then floats back down again in a movement that travels around the stadium. Yet, viewed from the individual seat, all you can see is an up and down movement from the individual sitting there. The people fixed in their seats create a longitudinal movement (horizontal) around the stadium by a series of singular vertical movements.

It appears as if the row of people run around the stadium whilst in fact they remain in their seats. Nobody moves away from his or her place, although it appears as if they do from an outside perspective. What runs around the stadium is the **Idea** itself. What gets displaced is the up-and-down movement of the individual. That is what is passed on to the person sitting next to the first one, and it is just the passing on of that information that makes it appear to an outside observer as though the person moves around the stadium. The matter only vibrates vertically, but the way each piece of matter relates its vibration to the vibrations of each piece of matter that surrounds it, makes it appear as if a wave moves across space. The wave itself contains the information to which each particle it meets responds.

The information, in this case of "standing up and sitting down", is passed from person to person while nobody moves away from his or her place. By moving up and down and passing that information on to the next person, a wave is created that "moves around the stadium". The information that is passed on includes how high the person has to rise (amplitude) as well as how often they need to do this per time unit (frequency). And there you have your wave.

In the same way, all other waves are created. Hence, the water molecules in the sea only move up and down – oscillate – but do not travel in any other direction. In other words, the wave that we observe moving the sea up and down is not caused by water molecules running away, but is caused by the passing on of the information of oscillation from one molecule to the next. This creates the impression, from the outside, of the movement of the water units while, in fact, all that moves is the up-and-down movement of the molecules themselves in a coordinated and synchronised way.

Gusts of wind "moving" through the air are, in essence, a passing on of the information of oscillation from one point to the next. The air molecules remain in place, just dance up and down, and create the illusion of wind movement through the air. All that moves through the air, however, is **information**.

Light is a wave, which vibrates perpendicular to the direction of its motion. For example, if a light is moving towards you, its vibration can be in the vertical or horizontal direction. This has practical use in polarised sunglasses, for example, which reduce glare by blocking all light that vibrates in the non-chosen direction. So, quantum physics already knows that the movement of a wave is perpendicular to its vibration. This is exactly what we are seeing in the material world in the example of the Mexican wave. One force seems to be pushing the wave in one direction, creating a vibration in another plane.

One of the most important principles of the quantum theory – the Heisenberg Uncertainty Principle – states that there is a finite probability that seemingly implausible events can happen. For example, the possibility of finding on Jupiter one particular electron identified on earth is very real, but very unlikely. The likelihood of finding the electron anywhere increases dramatically according to its energetic state. Electrons at low energy behave very much like a wave, while high-energy electrons behave like point-like particles. Because of this dual nature of electrons, quantum physics postulates that matter exhibits both wave-like and particle-like characteristics. The quantum theory says that the electron is a point particle, but the probability of finding it is given by the square of the Schrödinger wave function. As the electron speeds up, the wavelength of the Schrödinger wave gets shorter, so the probability of finding it peaks around a singular point. As the electron slows down, the wavelength expands, and the probability of finding it expands out over space. As a result, we cannot precisely locate the position **and** velocity of the electron at the same time (The Uncertainty Principle). It is the power of the force perpendicular to the plane in which a point in the universe is vibrating that determines the speed of vibration of the electron (universe point).

Between them, the strong force and the weak force "create" the matter point. It's the interplay between them that organises the density that we recognise as "matter". These forces act within one plane as they are opposing forces, so we can imagine them moving a point in the universe up and down. It is this local vibratory movement that determines the kind of matter "point A" will display. If you see this point moving in space it is because of the two other forces interplaying – the electromagnetic force and gravity – again working in opposite directions but within another plane. This plane is perpendicular to the plane in which the strong and weak forces operate. It is the impulse from the resulting electromagnetic and gravity force that stimulates "point A" in the universe to show up as a particular point of matter, by moving it up and down in a very specific way with very specific vibratory characteristics.

As the electromagnetic force moves along in the plane perpendicular to the "matter-formation", it stimulates the next point, "point B", in a similar way, which has the effect of inducing "point B" to become the same matter point as we have seen in "point A", which in turn has now lost the impulse to show up as that particular matter. It appears to an observer, who is separate from the point itself, as if matter has moved from "point A" to "point B". If the stimulating force hitting any point is relatively weak, all we can observe is a wave-like function through the ether. If, on the other hand, this force creates a high vibration level at that particular point in the universe, we will recognise it as a point of matter; there has been enough energy transfer from wave to point for it to have a density that we recognise as a point.

What determines that particular place of the universe to be "air" or "water" is just the information that the strong and weak forces bring to that point inside the matter-forming plane; in other words the type of high-frequency vibration of the point determines what "kind" of matter we see. We know that each atom has its own vibration pattern; each physical particle has its own specific vibration frequency, which is a resonance frequency from all forces influencing that point, resulting in a movement in this particular direction. We also know that each particle is actually a wave, unless we view it in a specific way in which we "fix" it in a position whereby it becomes **physical**. Now it is viewed as a particle, which has specific physical properties, such as weight, dimensions, electric charge, spin direction etc. All these properties are, in fact, nothing more than manifestations of the wave pattern that was there before we "froze" it in time and started seeing it as a particle. The information that the wave contained has been **solidified** by our observation into physical properties. This information is held within the weak and strong forces. These distinct properties have been given different names so that we can relate to our observation of the expression of information. One we have called "water", and the other we have called "air".

Every point of the universe has the potential to vibrate with the information that is within the electromagnetic spectrum to "create" the atoms of hydrogen, oxygen, helium and all others. These atoms are only the physical manifestation of the information contained in the oscillatory pattern of the universal point. Every point can then quite easily, and in rapid succession, change from one atom creation to another by altering its vibration. Matter is thus created by each point of the universe at each passing moment of time. Or, in other words, viewed from the static point of the universe, or the single seat in the stadium, it has the potential to be all particles – the person in the seat can be anywhere between fully up and fully down (which is the effect the full electromagnetic spectrum can have on that particular point). In one plane, matter is created by the weak and strong forces; in another plane, at right angles to the first one, the apparent movement of the particle is created by the electromagnetic force and gravity.

It was Einstein who first showed what energy really is and that matter is nothing but trapped energy. When energy congeals into particles of matter it produces a negative imprint in the energy field. Also, in his theory of relativity, Einstein showed that the mass of a body gets larger and larger the faster it travels. This process speeds up the closer the mass travels to the speed of light. The object becomes extremely large and, as one tries to reach the speed of light, the mass becomes infinite, which increases the resistance to acceleration to the point where it becomes impossible to move the object. The only thing that can travel at the speed of light has to be a thing without mass, such as light itself!

Even a stationary object contains energy, which is locked within its mass within its constituent atoms. The amount of this energy is the amount of mass (m) multiplied by the speed of light to the power of 2. The famous equation as we know it: $E = Mc^2$. In order to adhere to the law that says that there cannot be any loss of energy, only a transformation of energy, we need to include the kinetic energy of a moving body. You might be inclined to simply add the kinetic energy to the energy contained in the mass (mc^2). This would be true but for the fact that when in motion, the object's mass increases and therefore mc^2 will change too. You calculate this by first adding the square of the energy of motion to the square of the energy in its mass, and taking the square root of this sum to give the answer of the total amount of energy of a moving body:

$$E^2 = (Mc^2)^2 + (Pc)^2 \text{ with } P \text{ representing the momentum}$$

$$E = Mc^2 + \sqrt{(Pc)^2}$$

The implications are astonishing. First, massive objects at rest contain an amount of energy mc^2 within them. Second, even something that has no mass, such as a photon of light travelling at the speed of light, will have energy due to its motion. With the law of conservation of energy, it is therefore perfectly possible for the energy in a beam of light to become trapped within matter. Or, in other words, the energy travelling as an electromagnetic wave is being transformed into matter, thereby losing motion energy. The energy of a photon of light (which has no mass and travels at the speed of light) becomes trapped in a small particle with a negative electric charge – the electron – and its complementary piece of substance – the positron. The positron is a "virtual particle" or antimatter.

Here we can get the first idea of how our universe emerged from the Big Bang. The initial enormous amounts of heat and light started to congeal into counterbalanced pieces of matter and antimatter. This is suggested by Einstein's theory of relativity, with the implication that matter leaves an invisible imprint on the fabric of the universe, a kind of "mirror image" of the matter itself, called antimatter. While relativity explains the various energy accounts, it is when relativity is combined with quantum mechanics that the full power of nature is revealed.

If everything in our universe has come from light, it is about time for us to understand light a bit better.

Chapter 2
Light and the Golden Ratio

As mentioned in the preface, I discovered an awful lot of information about the Great Pyramid through my investigations of literature on the subject: measurements, relations, slices, orientation and more. What struck me was how many people have researched the why and how of this phenomenal monument. Still, after all my reading, I was left with a feeling of dissatisfaction. I kept coming back to words like "Ta Khut" and "Rostau". The Egyptians called the pyramid "Ta Khut" and Gizeh was known as "Rostau". Ta Khut means **The Light** and Rostau means **The Gateway to Another World.**

Questions that arose for me from words like "light" and "gateway" were:

- What is light?
- Where does light come from?
- What is the purpose of this gateway?
- Where does the gateway lead to?
- How do you construct a gateway?

In considering this question, we may quickly decide that *Light Is an Electromagnetic Wave Within a Frequency Range That Is Observable to the Human Eye.* And then the investigation starts! You carry on asking "why", just like small children do. As adults we have rejected this kind of questioning because we find it boring, but also because we run out of answers after a while. The series of questions and answers then tend to end in just "because".

Let's ask ourselves the next question: "What is an electromagnetic wave?" Electromagnetic waves are electrical and magnetic vibrations that travel through space. Light is one such form of electromagnetic wave that moves along, carried by particles called photons. The speed of light is around 300,000 km/sec.

The next step is the electromagnetic spectrum, in which you will find the division of electromagnetic waves according to their frequency. We go from radio waves to radar, to microwaves, to infrared, to visible light, to x-rays, gamma rays and cosmic rays.

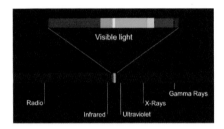

Here, we concentrate on visible light. The visible spectrum of light has a known wavelength of between 780 nm and 380 nm. A bundle of light is composed of several bands of wavelengths, which manifest to us as different colours. By way of a prism, "white" light can be broken down into the different colours that make up light. The main colours that occur this way are red, orange, yellow, green, blue, indigo and violet. These are the colours of the rainbow, and which also occur as a result of breaking light up as it passes through water droplets.

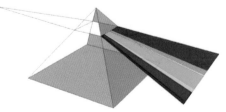

Just as a musical note carries within itself all other notes, so it is with colours, whereby every colour has all of the other colours within it. What we see as red within the spectrum is in fact a combination of red, orange, yellow, green, blue, indigo and violet. However, all these colours have to be combined in a very particular way in order to produce the red colour. Can we examine the composition of every colour and find out what the combinations are? We can and we will.

Visible Spectrum

The visible spectrum is that part of the electromagnetic spectrum that can be seen by the human eye. That is why we call the electromagnetic waves lying within this spectrum the "visible light".

The visible spectrum has a wavelength of between 380 nm and 780 nm in vacuum, which corresponds with frequencies of between 400 and 790 terahertz. The eye perceives the various wavelengths as different colours: red for the longest wavelength and violet for the shortest. The highest sensitivity of the human eye lies around 555 nm (yellow–green) during the day and around 507 nm (blue–green) at night. However, not all colours that the human brain can distinguish are manifested within the visible light spectrum; these can only be achieved by mixing two colours together.

The oldest reference to the seven base colours of the visible light spectrum can be found in the writings of Samba Purana from old India, dating back to 1500 BC.

Isaac Newton was the first to coin the term "spectrum" and to write about his findings in his book *Optics*. He used the Latin word spectrum to describe the set of colours that appeared when he filtered sunlight through a glass prism. This showed the colours of the rainbow, as mentioned earlier, in ever-decreasing wavelengths (increasing frequencies). Beyond visible light, the spectrum continues in ultraviolet, x-rays and so on in ever-increasing frequencies, and via infrared and microwaves to radio waves in ever-decreasing frequencies.

Dividing the spectrum is usually done using wavelengths, meaning the wavelengths in vacuum because the wavelength is dependent upon the medium the wave travels through. The advantage of using frequencies is the fact that they remain the same regardless of the medium the wave travels through. However, when describing the bending, the interference and the dispersion of waves in a particular medium, wavelength is the most relevant entity.

Light that consists of waves that have the same wavelength/frequency is called monochromatic light. The colour that manifests is the colour that directly relates to that frequency. In nature, almost all light is polychromatic, consisting of a variety of frequencies. To the eye this manifests as one colour too, but this is the result of the combination of the various monochromatic frequencies within the one wave. When the various frequencies are combined in a particular way, we perceive the light as white. Other colours can only manifest in specific colour combinations, like brown, for instance. The various colours of the visible light spectrum have their own band.

The visible light spectrum is composed of:

32.50% red	between	780 nm	and	650 nm
16.25% orange	between	650 nm	and	585 nm
2.50% yellow	between	585 nm	and	575 nm
21.25% green	between	575 nm	and	490 nm
17.50% blue	between	490 nm	and	420 nm
10.00% violet	between	420 nm	and	380 nm

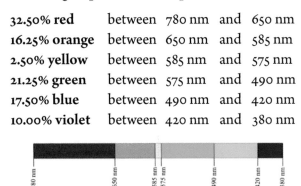

Research has shown that the various colours appear with different intensity depending on the light source. The spectrum emanating from the light source is a kind of signature. From the light radiating from a hot gas, for example, we can deduct the precise chemical composition of the gas itself. In a similar way, the spectrum of a hot solid material indicates the temperature of the material.

The word spectrum is used in a much wider sense to indicate the intensity of any kind of wave phenomenon. The spectrum now entails the description of the wavelengths and their strengths as they are present within the wave signal. This can be visualised in a graph in which the vertical shows the amplitude and the horizontal the frequency.

For sound waves we talk about a sound spectrum, which tells us about the high and the low notes contained within the sound. For visible light, the spectrum indicates the colours that are present within the light.

A spectrum can be a full spectrum, which means that all wavelengths are present in a certain wave. For instance, a light bulb has a full spectrum in which all wavelengths appear in more or lesser form. The alternative is a line spectrum in which only a few wavelengths dominate the spectrum. The spectrum of a sodium lamp – the classic yellow–orange street lighting – has a line spectrum, and so have the TL lights.

A combination of a full spectrum and a line spectrum we can see when a full spectrum falls on a material that absorbs specific lines. This results in a full spectrum where the specific lines are missing. This kind of spectrum is useful in atom absorption spectrometry as well as in astronomy, where stars are classified in groups according to their spectrum.

As far as sound is concerned, "noise" is a typical full spectrum, and "musical notes" are a typical line spectrum. A musical note created on an instrument consists of a base note, or base frequency, and a number of overtones. These have a frequency that is a multiple of the base tone, but the energy of the overtones is inversely proportional to the frequency. This is called "timbre", which is a combination of base and overtones plus a variety of beat frequencies. Speech is a combination of notes (vowels) and noise (consonants).

Light is the electromagnetic wave with a frequency range that can be seen by the human eye, added to which we have the infrared range – a slightly lower frequency – and the ultraviolet – a slightly higher range. The three variables that describe light are the brightness (amplitude), the colour (wavelengths), and the polarisation (direction of vibration), which is always perpendicular to the direction in which the wave moves. Optics is the study of light and the interaction between light and matter.

What we have provided here is a short summary of the information that is available on the visible spectrum. When researching this, what struck me was that there was no explanation about the specific frequencies of the spectrum and their meaning. Ian Gurney, in his book *The Cassandra Prophecy*, talks in Chapter 3 about "the beast". And in the Bible, The Book of Revelation 13:18, it reads: "*This Calls for Wisdom: Let the One Who Has Understanding Calculate the Number of the Beast, for It Is the Number of a Man, and His Number Is 666.*"

Ian maintains that many interpretations have been given for the number 666. Some have linked it to the antichrist itself. Many have used numerology, codes and cryptic anagrams to attempt to solve the mystery. Sometimes, theology loses sight of the fact that The Book of Revelation was a simple, uncomplicated message, given to a simple, uncomplicated man in the first century AD. Although John was no doubt an intelligent and deeply religious man, he was not a complicated man, which allows us to dismiss all intricate and difficult theories surrounding the number. There is a much simpler solution, which was immediately recognisable to John too, who was well educated in the religious teachings of his time; Christian as well as Jewish.

The number six is indeed a figure of a man, but not of one man. "*So God created man in His own image … And the morning and the evening were the sixth day.*" (*Genesis 1:27/31*).

On the sixth day, God created mankind and blew life into him. The number six stands for all mankind, the human race, and three times the number six is a human being trying to emulate God, the three in one – Father, Son and Holy Spirit. The number 666 is an expression of the unholy trinity, the person who turns away from God and views himself as almighty, using money and greed as his gods, and power and supremacy as his goals.

Humanity has reached a crossroads. We haven't concerned ourselves with what created us but instead we have considered ourselves creator, which is our fundamental mistake. For centuries people have been trying to understand the meaning of the beast without reaching the simplest of conclusions – we are the beast!

Whether Ian is right or not, I will leave the question open for now and we will return to the subject later on. Gathering information through reading or listening to someone else is a way of receiving input, which you can interpret or simply ignore. However, it is my belief that everything you do, everything that happens in your life, is no coincidence. What I have concluded from Ian's explanation is that we, as the whole of humanity, have forgotten something important: we are not almighty and life should have more meaning than money, greed, power and supremacy. It appears that we do not know the meaning of indigo. I believe that if we manage to give six (indigo) a place within the light spectrum, we have made a huge step forward.

You may wonder why I put so much emphasis on the number six and why Ian Gurney's writings are so important to me. Well, I hear everywhere that we need to take control of our own lives. For me that started when I got the message that we can lead our own life, in a literal sense; you can send it in a specific direction. Someone said, a long time ago, *"Ask and You Will Be Given."* One-hundred years ago, Einstein put those same words into a mathematical formula: $E = mc^2$.

Energy equals mass multiplied with the speed of light squared. To me this means: ask and you will be given. Maybe you will have to wait a little while, but you will get the answer. But how will you get it? Life offers you these answers in what we consider to be coincidences. What it requires from us is that we understand the meaning of these coincidences, these answers. For this to happen we need to learn to use our senses properly. If, for instance, you have been thinking a lot about someone you haven't heard from in ages, the chances are that you will bump into that person or that the phone will ring with that person on the line. You have "created" the coincidence!

To me, coincidences do not exist. Coincidences are answers and life can be guided by questions and answers. I believe this is true for all of us on this planet. We all help to create our own reality. To do so, we have to learn to ask the relevant questions in order to get the relevant responses.

The Golden Ratio

The Golden Ratio, also known as "divina proportione" (the divine proportion), looks like an ordinary number; however, when you dig deeper you realise that this simple number returns in many formats. For instance, it is an essential part of nature. People use it often, either consciously or not. It plays an important part in art and architecture, for example. The first person that we know of who studied the Golden Ratio was the famous mathematician, Pythagoras.

The Golden Ratio is the way the length of a line is divided in a very special manner. The largest piece of the line relates to the smaller part as the whole line relates to the largest piece.

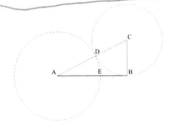

In order to find the piece E on the line AB, we draw BC perpendicular to AB, so that BC = ½ AB. Then we draw a circle with point A at the centre and AD being the radius (BC = DC and AD = AC − DC). This last circle crosses AB in E.

The mathematical formula for the Golden Ratio is $\phi = (1 + \sqrt{5}) / 2$. ($\phi$ is the Greek letter phi).

So, in this example:

$$B = A\,(\text{-}1 + \sqrt{5}) / 2$$
$$B = A\,(\text{-}1 + 2.236067977\ldots) / 2$$

When A equals 1 than B equals 1 $(\text{-}1 + \sqrt{5}) / 2$

$$B = 1(\text{-}1 + 2.236067977\ldots) / 2$$
$$B = 1.236067977 / 2$$
$$B = 0.618033988\ldots$$
$$A − B = 1 − 0.618033988\ldots = 0.38196601\ldots$$

The Golden Ratio and the Visible Light

When you use the Golden Ratio on a line, you divide the line in a very particular way. The rule is that:

$$X = A + B + C + D + E + F + G \text{ (seven pieces)}$$
whereby A = B + C, B = C + D, C = D + E, D = E + F, and E = F + G.

Having used the Golden Ratio, it turns out that the composition of:

$$A + B + C + D + E + F + G \text{ does not equal } X.$$
There remains a leftover R.
$$X = A + B + C + D + E + F + G + R$$

The above does not comply with the rules. Apparently it is impossible to divide a said line according to these rules.

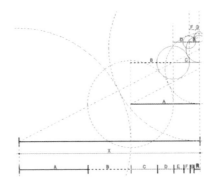

I still want to achieve this. The question is: "how can it be done?"
Let's look at it from a different perspective.

What happens if we turn it around and start with a line A.
Using the Golden Ratio, we can now divide this line into B and C,
whereby A = B + C. Then we divide B into C and D, whereby B = C
+ D. We keep repeating this method until we have seven pieces of
the following nature: A = B + C, B = C + D, C = D + E, D = E + F and
E = F + G. In this series, A is the largest piece and G the smallest.

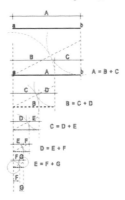

Now put these pieces together, join them up to form one line
and we do have a line that corresponds to X = A + B + C + D + E
+ F + G.

A = B + C, B = C + D, C = D + E, D = E + F EN E = F + G.

From the light spectrum we know that yellow is the smallest piece with a wavelength between 585 nm and 575 nm, and we know that 580 nm is the middle of the light spectrum wavelengths. If we now place our smallest piece, G, in the middle part of line x, then we have several options to try to place the other pieces within the line. Looking at the light spectrum we know that red and orange come before yellow, and only two pieces can be used to fit onto the line in those positions: A and D. That leaves us with B, E, C, and F. Now we have, of course, several possibilities. Working out the variations, it turns out that only one combination satisfies all requirements (we go into detail on this later) and that is A–D–G–B–E–C–F. Put from smallest to largest, we get: G–F–E–D–C–B–A.

If we now give A the number 1 and colour it red, D the number 2 and orange, G becomes 3 and yellow, B becomes 4 and green, E becomes 5 and blue, C becomes 6 and indigo and F becomes 7 and violet, then we have the spectrum of visible light according to the Golden Ratio, showing the proposed seven colours as expected.

The earlier code, A–D–G–B–E–C–F, now becomes 1–2–3–4–5–6–7. This is the manner and order in which the light spectrum presents itself to us (the rainbow). The code G–F–E–D–C–B–A becomes 3–7–5–2–6–4–1, in which 3 is the smallest and 1 the largest part of the spectrum. This is the order in which the spectrum is formed out of white light.

The visible spectrum, divided according to the Golden Ratio and containing seven colours, consists of:

39.55907%	red	between 780.0 nm and 621.7 nm
9.33863%	orange	between 621.7 nm and 584.4 nm
2.20455%	yellow	between 584.4 nm and 575.5 nm
24.44880%	green	between 575.5 nm and 477.7 nm
5.77156%	blue	between 477.7 nm and 454.7 nm
15.11025%	indigo	between 454.7 nm and 394.2 nm
3.56707%	violet	between 394.2 nm and 380.0 nm

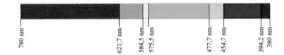

Ordered from the smallest to the largest piece, the series becomes: yellow (3), violet (7), blue (5), orange (2), indigo (6), green (4) and red (1).

The whole of creation inside our universe comes from the light spectrum and follows the laws of this creation. **Everything Is Made in the Same Way.**

According to Ayurveda, the seven types of tissues also appear in a very specific way. From the seed, the first tissue that is created is juices (plasma), then blood, then muscle, then fat, then bone and lastly nerve tissue (more on this later) to return to seed again. If the creation of the tissues is following the same rules as the way the colours of the rainbow are created – and why shouldn't it, as everything is being created from light? – we create the tissues according to the pattern.

The light spectrum itself varies from red to violet, and to make it easier we will number the colours that way: red 1, orange 2, yellow 3, green 4, blue 5, indigo 6, and violet 7. As we know from the yellow part of the spectrum, all the other colours are being created in a particular order and according to the Golden Ratio, then there are only two possibilities that need to be explored.

We can look at the sequence from smallest to largest, which results in: from yellow violet appears first, then blue, then orange, then indigo, then green, and finally red. This leaves us with the code: 3–7–5–2–6–4–1.

Or we can turn it around: from yellow appears red first, then green, then indigo, then orange, then blue and finally violet. This leaves us with the code: 3–1–4–6–2–5–7.

One of these two is the origin code for matter. So which one is it?

Just as every musical note contains within itself the other notes, we find that every colour consists of all the other colours. What we perceive as red within the visible spectrum is in fact a combination of red, orange, yellow, green, blue, indigo and violet. In order to show up as a particular colour, the mix of all colours must be in certain proportions to make the overall differences of red, orange and so on. Let's have a look at these particular combinations.

To start with, we know from Einstein's formula, $E = mc^2$, that E is energy, m is mass and c is the speed of light. Hence, everything we perceive in the material world is a perception of energy. From quantum physics we know that subatomic particles of matter are very abstract entities with a double nature. Depending on how we look at them, they sometimes behave like a particle and sometimes like a wave. Light also has this double nature, because it is an electromagnetic wave, but it can also appear in the form of photons, which are light particles. This is difficult for us to visualise and accept that something is, at the same time, a particle *And* a wave, because a particle is an entity with a mass and a specific place in space, and a wave has no mass and is spread out over a large area of space. We could say that the mass part is an expression of energy, but that the energy doesn't necessarily need a carrier to exist.

When we look around us we see a great variety of objects, which are all an expression – a visualisation – of energy. Plants, trees, animals and humans are all expressions of energy. We see them and recognise them because of their shape, movement and other characteristics. Various types of dogs are recognised as dogs but are also recognised as not being the same kind of dogs. Energy expresses itself in a great variety of forms, structures, patterns and mass shapes. This means that those differences must also be found within the energy that has produced the specific mass shape. Therefore, every single shape presented in matter must have a corresponding energy pattern, which leads us to conclude:

- All matter has a double nature: particle form (mass) and electromagnetic radiation (energy).
- All matter has two forms in which it presents itself: a mass form Mp and an energy form Ep.
- Human beings have a double nature: a body, which is the mass form Mp (physical world) and a mind, which is the energy form Ep (metaphysical world).

Referring once more to Einstein's formula of $E = mc^2$, in which E represents energy, m is mass and c is the speed of light, we can deduct that $Ep = Mp.c^2$, or in words: the energy pattern Ep equals the mass pattern Mp times the speed of light squared. This implies that when the energy pattern changes, in time, the mass pattern will change too. In other words, **Everything Evolves**.

Let's take another look at the formula:

$$Ep^2 = (Mp.c^2)^2 + (Pc)^2 \text{ or } Ep = Mp.c^2 + \sqrt{(Pc)^2}$$

This means that when the energy pattern changes, in the long run, the mass pattern changes too, but also that the direct environment of the mass alters, which is here represented by $\sqrt{(Pc)^2}$.

Chakras and the Human Energy Pattern

Chakra is a Sanskrit word that means "wheel". A chakra is described as a wheel-like vortex in which various levels of consciousness meet.

Through the rotation of the chakra there is a constant movement of energy, either inwardly or outwardly; a constant exchange of energy with the cosmos. The chakras are receivers, transformers and distributors of *Prana*, which is life energy. It is through their function that we exist. When a chakra is not functioning well, as in it is taking in too little energy or giving out too much energy, an imbalance will occur between the chakras that can lead to illness. This could manifest itself in a body area or organ where the chakra is prominently present. In total there are seven main chakras situated from the pubic bone to the crown. These are the elementary energy centres, which are like gateways between the human energy field and the physical manifestations within that field.

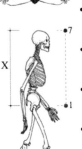

- The first chakra is about survival and is located at the base of the spine.
- The second chakra is about sexuality and desire and is located in the lower abdomen.
- The third chakra is about personal power and energy and is located by the stomach.
- The fourth chakra is about balance and is located by the heart.
- The fifth chakra is about communication and is located at the throat.
- The sixth chakra is about imagination and intuition, and is located between the eyebrows.
- The seventh chakra is about insight and knowledge and is located at the crown.

The seven chakras relate to the seven colours of the visible light spectrum, as presented in ancient texts, and to the whole of the electromagnetic spectrum itself. The chakras are presented in the same way: 7–6–5–4–3–2–1, whereby 7 corresponds to the colour violet, 6 to indigo, 5 to blue, 4 to green, 3 to yellow, 2 to orange and 1 to red.

Applying the Golden Ratio, we can begin to visualise this

information. We know that chakra 1 is situated right at the base of the spine and chakra 7 right on the top of the head. The distance in any human being between those points is the length of the spinal column plus the height of the head. Let's call this distance x. This distance is measurable by sitting up against a wall with a straight back. x is then the distance between the floor and the top of the head.

This locates chakras 1 and 7.

From this information, and dividing this line according to the Golden Ratio, we can now deduce the location of chakra 3.

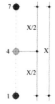

Chakra 4 is placed in the middle of x. (Explained shortly.)

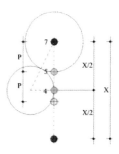

Then we can place chakra 5 between 4 and 7 according to the principle of the Golden Ratio.

Notice that chakra 5 is exactly halfway between the position of chakra 3 and that of 7, as chakra 4 is halfway between 1 and 7.

Chakra 2 is placed between 1 and 3 according to the Golden Ratio.

Chakra 6 is at the point between 5 and 7 according to the Golden Ratio.

You may have wondered why we arbitrarily placed the fourth chakra midway between 1 and 7. Our choice is now confirmed in the next drawing. Chakra 4 is at the exact point between chakras 2 and 5 according to the Golden Ratio.

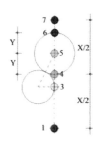

We further note that the distance between 4 and 5 is equal to the distance between 5 and 6.

Putting it all together we get the following:

These are the positions of the chakras as created by the process of drawing the information. The beauty of this is that if you can draw something, you can also calculate it. This means that for a crown to base length x we can calculate exactly where the chakras are for every measurement that exists.

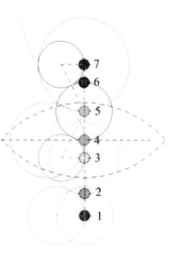

For example, for a person who is 169 cm tall and has a crown-to-base length of 87 cm, this means that, standing in the upright position, chakra 7 is 169 cm off the floor, and the first chakra is 82 cm off the floor. The total length, 169 cm, minus the distance between the floor and the first chakra is then the crown-to-base length.

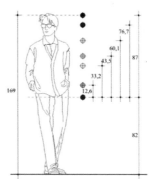

Having drawn the energy pattern of a person, it is time to experiment a bit more. The visible light spectrum shows itself as violet 7, indigo 6, blue 5, green 4, yellow 3, orange 2, and red 1. When we look at the energy pattern of a person in a similar fashion, it is created as follows:

● 7
● 6
◑ 5
◑ 4
◈ 3
◑ 2
● 1

Following the principle of the Golden Ratio, we know that in the visible light spectrum the composition is as follows:

7		3,56707 %
6		15,11025 %
5		5,77156 %
4		24,44888 %
3		2,20455 %
2		9,33863 %
1		39,55906 %
		100 %

These proportional divisions will recur time and time again, only in different combinations.

The Creation Code of the Visible Spectrum

Composition of YELLOW

The light spectrum itself varies from red to violet. We are going to use the numbering system to make it easier, in that red 1, orange 2, yellow 3, green 4, blue 5, indigo 6, and violet 7. We know that the spectrum starts with yellow and follows a specific order, according to the Golden Ratio. Now we are going to investigate the two possibilities, which are: from smallest to largest, or from largest to smallest.

First, we investigate the series from smallest to largest, which is: yellow, violet, blue, orange, indigo, green and finally red. This results in a numerical code of 3–7–5–2–6–4–1. The question now is whether this is the creation code of the visible light spectrum.

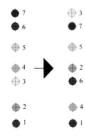 On the left-hand side in this diagram, we see the chakras presented as we have placed them according to the Golden Ratio. On the right-hand side we see the place of the chakras numbered differently according to their appearance in creation. The smallest part in creation, 3, is placed on the actual measured smallest part of the spectrum, which is position 7. The rest follows on from there. Now, follow the creation code as you place the figures on the corresponding chakra places as you move down the list. So: 3 on 7, 7 on 6, 5 on 5, 2 on 4, 6 on 3, 4 on 2, and 1 on 1.

If we then draw circles with 3 (see below), which has been placed on position 7, as the centre point, we can visualise the distances between each number and the number three. Enlarging the radius as we move along, we first meet the colour violet, then blue, then orange, then indigo, then green and finally red, reflecting the code we have used.

Putting this into figures gives us the following code: 3–7–5–2–6–4–1.

With this code it is now possible to calculate the proportions of the various colours within yellow and, using the Golden Ratio, we can draw the results. As red is created much later within the yellow spectrum than, for instance, violet, its contribution to the colour yellow is far less significant. YELLOW is mostly made up of yellow while red plays the least part, according to the code 3–7–5–2–6–4–1. This is how the yellow colour comes into existence.

The composition of the yellow spectrum is as follows:

3		39,55906 %
7		24,44888 %
5		15,11025 %
2		9,33863 %
6		5,77156 %
4		3,56707 %
1		2,20455 %

In yellow we find mostly yellow, then violet, then blue, then orange, then indigo, then green and red, which is the least present.

Let's check this. To give continuity to the complete visible spectrum, the middle part, 580 nm, should stay in the middle with every further division that happens according to the Golden Ratio principle (39.55906% yellow, 24.44888% violet, 15.11025% blue, and so on). The flow through the centre must be guaranteed.

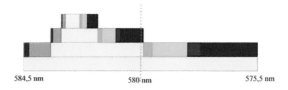

584,5 nm   580 nm   575,5 nm

The picture above shows that the composition of yellow according to the code 3–7–5–2–6–4–1 is not the right one. From our knowledge of colours, we also know that if we mix these colours in these ratios we will never end up with a yellow colour. The combination of these percentages is not correct.

That leaves us with possibility two, code **3–1–4–6–2–5–7**. Here, we code from the seed colour 3, following the development from largest to smallest.

Comparing 3 with a seed, it follows that the other colours appear in order, ending up with the very smallest of them all, again being the number 3, which we mark as 3′ to avoid confusion.

A seed germinates and creates a plant, or an animal, or a human being. These grown plants, animals or humans are all capable of producing seeds to perpetuate the race and species. That is how evolution progresses. We complete the code now as 3–1–4–6–2–5–7–3′, whereby 3′ represents the seed. We could say that 3 has evolved into 3′. This results in the following:

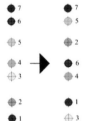 On the left we see the known places of the chakras according to the Golden Ratio. On the right we have the numbers placed on these chakra positions starting with the seed. This time we begin at the bottom because we place them from largest to smallest, and the red colour, place 1 on the base chakra, has the highest contribution to the visible light spectrum. So we get 3 on 1, 1 on 2, 4 on 3, 6 on 4, 2 on 5, 5 on 6, and 7 on 7. Again we can draw circles with number 3 as the centre.

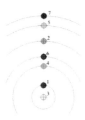

We now notice that red is closest to yellow, then green, then indigo, then orange, then blue and then violet. Now the code that gives us the influences within the yellow is 3–1–4–6–2–5–7, which gives us the following composition of yellow:

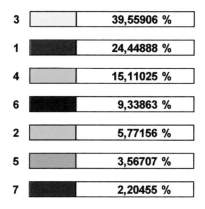

Shown as a drawing:

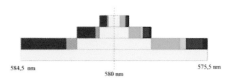

We can now see that the centre of the visible spectrum, 580 nm, remains at the centre of yellow by every further division we care to operate. The continuity of the system is hereby guaranteed. Hence, code 3–1–4–6–2–5–7 is the creation code of matter, visible light being the first "matter" to appear in creation.

From this configuration we can determine the composition of all the other colours within the visible spectrum.

Composition of RED

From chakra 1 we can draw circles reaching the other chakras.

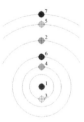

This produces the code 1–3–4–6–2–5–7 for red. This produces, according to the Golden Ratio, the following internal proportions:

1		**39,55906 %**
3		**24,44888 %**
4		**15,11025 %**
6		**9,33863 %**
2		**5,77156 %**
5		**3,56707 %**
7		**2,20455 %**

Shown as a drawing:

780 nm   621,7 nm

Composition of GREEN

From chakra 4 we can draw circles reaching the other chakras.

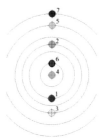

This produces the code 4–6–1–2–3–5–7 for green. This produces, according to the Golden Ratio, the following internal proportions:

4	39,55906 %
6	24,44888 %
1	15,11025 %
2	9,33863 %
3	5,77156 %
5	3,56707 %
7	2,20455 %

Shown as a drawing:

575,5 nm   477,7 nm

Composition of INDIGO

From chakra 6 we can draw circles reaching the other chakras.

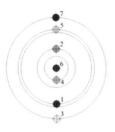

Here we notice that chakras 7 and 3 are equally distanced from chakra 6. This leaves us with two possibilities to compose indigo, and consequently we will have two possible codes:

INDIGO B with code 6–4–2–1–5–7–3

INDIGO A with code 6–4–2–1–5–3–7

INDIGO B with code **6–4–2–1–5–7–3** is composed as follows:

6	39,55906 %
4	24,44888 %
2	15,11025 %
1	9,33863 %
5	5,77156 %
7	3,56707 %
3	2,20455 %

Shown as a drawing:

454,7 nm   394,2 nm

INDIGO A with code 6–4–2–1–5–3–7 is composed as follows:

6	39,55906 %
4	24,44888 %
2	15,11025 %
1	9,33863 %
5	5,77156 %
3	3,56707 %
7	2,20455 %

Shown as a drawing:

454,7 nm   394,2 nm

Composition of ORANGE

From chakra 2 we can draw circles reaching the other chakras.

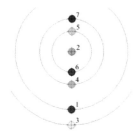

Here we notice that chakras 5 and 6 are equally distanced from chakra 2. We therefore have two choices as to which influence comes first and which second. Furthermore, chakras 7 and 4 are also equally distanced from chakra 2. Again, we have two possibilities, all of which results in four possible codes for the colour orange:

ORANGE A with code 2–6–5–4–7–1–3

ORANGE B with code 2–6–5–7–4–1–3
ORANGE C with code 2–5–6–4–7–1–3
ORANGE D with code 2–5–6–7–4–1–3

Therefore, orange can be composed in four different ways. Let's take a closer look at those possibilities.

ORANGE A with code 2–6–5–4–7–1–3 is composed as follows:

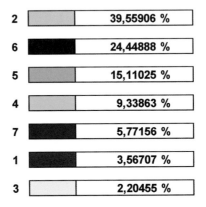

2	39,55906 %
6	24,44888 %
5	15,11025 %
4	9,33863 %
7	5,77156 %
1	3,56707 %
3	2,20455 %

Shown as a drawing:

621,7 nm   584,5 nm

ORANGE B with code 2–6–5–7–4–1–3 is composed as follows:

2	39,55906 %
6	24,44888 %
5	15,11025 %
7	9,33863 %
4	5,77156 %
1	3,56707 %
3	2,20455 %

Shown as a drawing:

621,7 nm   584,5 nm

ORANGE C with code 2–5–6–4–7–1–3 is composed as follows:

2		39,55906 %
5		24,44888 %
6		15,11025 %
4		9,33863 %
7		5,77156 %
1		3,56707 %
3		2,20455 %

Shown as a drawing:

621,7 nm   584,5 nm

ORANGE D with code 2–5–6–7–4–1–3 is composed as follows:

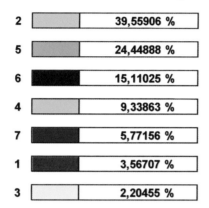

2		39,55906 %
5		24,44888 %
6		15,11025 %
7		9,33863 %
4		5,77156 %
1		3,56707 %
3		2,20455 %

Shown as a drawing:

621,7 nm   584,5 nm

Composition of BLUE

From chakra 5 we can draw circles reaching the other chakras.

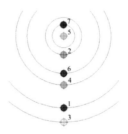

This produces the code 5–7–2–6–4–1–3 for blue. This produces, according to the Golden Ratio, the following internal proportions:

5		39,55906 %
7		24,44888 %
2		15,11025 %
6		9,33863 %
4		5,77156 %
1		3,56707 %
3		2,20455 %

Shown as a drawing:

477,7 nm   454,7 nm

Composition of VIOLET

From chakra 7, using this creative code, we can draw circles reaching the other chakras.

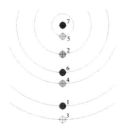

This produces the code 7–5–2–6–4–1–3 for violet. This produces, according to the Golden Ratio, the following internal proportions:

7		39,55906 %
5		24,44888 %
2		15,11025 %
6		9,33863 %
4		5,77156 %
1		3,56707 %
3		2,20455 %

VIOLET is composed mostly of violet, followed by blue, then orange, then indigo, then green, red and finally yellow.

Shown as a drawing:

394,2 nm   380 nm

Composition of the Colours of the Light Spectrum

We need to take the various codes into consideration when we compose the colours of the light spectrum. Because indigo can be composed in two different ways and orange in four different ways, this leads to eight possible ways to compose the light spectrum.

ORANGE A code 2–6–5–4–7–1–3 and **indigo A** code 6–4–2–1–5–3–7

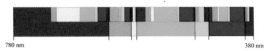

ORANGE A code 2–6–5–4–7–1–3 and **indigo B** code 6–4–2–1–5–7–3

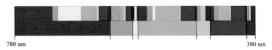

ORANGE B code 2–6–5–7–4–1–3 and **indigo A** code 6–4–2–1–5–3–7

ORANGE B code 2–6–5–7–4–1–3 and **indigo B** code 6–4–2–1–5–7–3

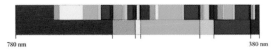

ORANGE C code 2–5–6–4–7–1–3 and **indigo A** code 6–4–2–1–5–3–7

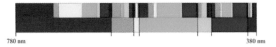

ORANGE C code 2–5–6–4–7–1–3 and **indigo B** code 6–4–2–1–5–7–3

ORANGE D code 2–5–6–7–4–1–3 and **indigo A** code 6–4–2–1–5–3–7

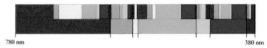

ORANGE D code 2–5–6–7–4–1–3 and **indigo B** code 6–4–2–1–5–7–3

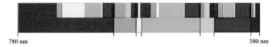

The question that remains is: "Why these eight different schemes?" The answer follows later.

Chapter 3
Visualisation of the Chakras

Further Division of Colours

RED

We have divided the colours into their seven colour patterns. Now it is time to go a step further and divide every colour again.

In the picture below we can see how the red part of the visible light, indicated by A, has been split according to the Golden Ratio into the seven component colours, as shown in B. Now we are going to do the same thing to all the colours in B according to the divisions we have found for each separate colour. Hence, the red in B is going to be split according to the colour code of red, which is 1–3–4–6–2–5–7. ORANGE in B is divided according to the code of orange, which is 2–5–6–7–4–1–3. This is the code 2D (which is appropriate in human beings) but the same thing can be done for 2A, 2B and 2C. YELLOW is divided according to the code 3–1–4–6–2–5–7; green according to 4–6–1–2–3–5–7; blue according to 5–7–2–6–4–1–3; indigo according to code 6B 6–4–2–1–5–7–3 (same for 6A); and violet according to 7–5–2–6–4–1–3.

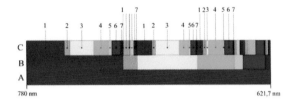

This further division is visualised in C. You can keep on dividing to eternity but the result is always the same: all colours contain all colours. From our research we now know that the representation C shows the locations where the various chakras are being expressed in the physical manifestation.

The following pictures show these C representations of the other colours of the spectrum.

ORANGE

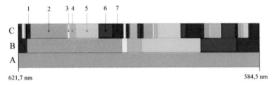

This is the orange code 2D (2–5–6–7–4–1–3) but the same can be done for 2A, 2B and 2C. INDIGO is represented as code 6B (6–4–2–1–5–7–3), as is done in the following pictures.

YELLOW

GREEN

BLUE

INDIGO

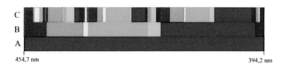

VIOLET

Placing these all together in their correct combination, we obtain the following picture:

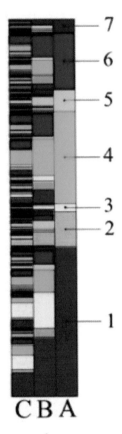

CBA

1 divided according to code 1–3–4–6–2–5–7
2 divided according to code 2–5–6–7–4–1–3 (code 2D)
3 divided according to code 3–1–4–6–2–5–7
4 divided according to code 4–6–1–2–3–5–7
5 divided according to code 5–7–2–6–4–1–3
6 divided according to code 6–4–2–1–5–7–3 (code 6B)
7 divided according to code 7–5–2–6–4–1–3

This gives a more detailed picture of the division of the visible spectrum. You may be asking, "Why do you want to do this?"

This all started one evening with a conversation about the research I was doing, more specifically about the human energy pattern, about everything being energy, some of which is converted into matter. It led to the statement that whenever there is a physical ailment there must be an energetic cause underlying it. This means that when my body expresses a certain pain in a specific place, the question arises as to where within the energetic pattern the fault can be found. Good question! But not so simple to answer. Driving home that evening, I realised that everything has come from the visible spectrum and so everything must be noticeable within that spectrum. The next day I started focussing my research on the visible light spectrum, which eventually, and much later, led to the above division. Let's have a look at what we can get from this division.

Visualisation of the Chakras

A chakra is nothing more than a communication gateway that fixes energy into matter. The human energy field has seven gateways that help to construct the physical body and exchange energy between the two manifestations. Let's explore this a little further.

I believe that the chakras are also ways of viewing the energy field from a variety of angles, not literally, but metaphorically. It helps to visualise a multidimensional field in a 3D format in which every chakra, or viewpoint, visualises a certain aspect of the field.

Let's start with the following picture, the division of the spectrum:

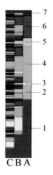

Let's then compare this with the human body, where the picture looked like this:

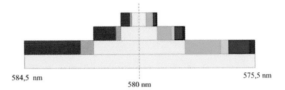

We can notice that the centre of the visible spectrum remains in yellow no matter how often we divide the colours.

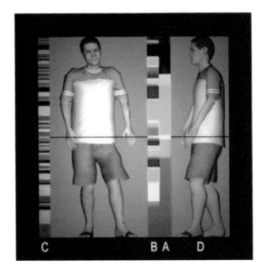

Here we have a visual picture of how the human energy field shows itself within the visible spectrum. Every small part of the human body can be created by using the same proportional division of every colour. This leads to an ever-increasing density within the wave function, leading to, eventually, the seven layers of matter.

In this picture, A stands for the visible light spectrum, B for the spectrum split into colours, representing the human field (we used 2D and 6B). In C we find the visualisation of the gateways of the chakras. You could also say that C represents the individual human field, as from this division the primary matter follows, from which a large specimen will be built, eventually. In D we show the composition of each colour, which refers to how much contribution each brings to each band.

We are now able to visualise the various chakras according to the calculations of the colours we made within the light spectrum. This shows us exactly where all the chakras are expressed within the physical structure.

Now we can compare the division of the spectrum of visual light with the actual physical body. Let's imagine lying down on a table or standing upright against a wall. By putting the division of the light spectrum next to the body we obtain the picture on the right.

We can now visualise the various change-over points according to the specific placement of all the colours within the light spectrum. It gives us a picture of where in life these lines manifest on the physical body.

The next picture shows us where we will encounter **chakra 1**. Because red is an integral part of every colour within the spectrum, it follows that chakra 1 manifests seven times within the red of the spectrum, seven times within the orange, seven in yellow, seven in green, in blue, in indigo and in violet. Therefore, the chakra expresses itself seven times in the visible spectrum.

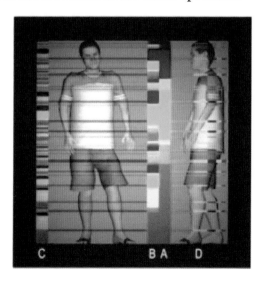

The next picture shows us where **chakra 2** expresses itself within the visible light spectrum. Again, we notice that chakra 2 manifests in seven times in seven places. It shows up seven times in the red part of the spectrum, seven times in orange, in yellow, in green, in blue, in indigo and in violet.

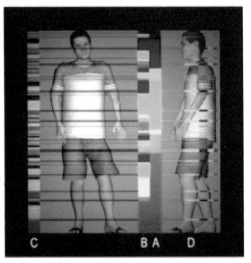

This picture is for **chakra 3**, and again it shows seven times in seven places, in the same way the previous ones did.

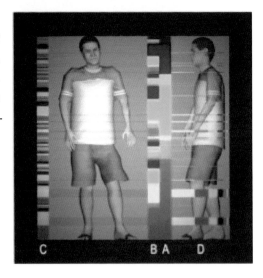

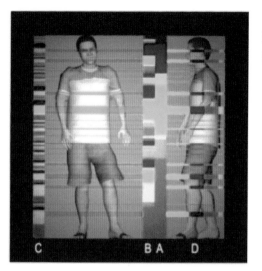

Next up is **chakra 4**.
Same thing again.

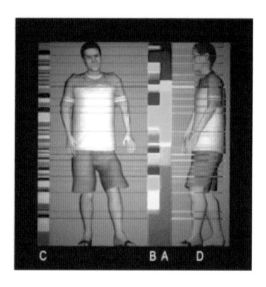

And chakra 5.

And **chakra 6**.

Here we can take note of the "third eye": indigo, indigo, indigo. This translates into the code as 6.6.6!

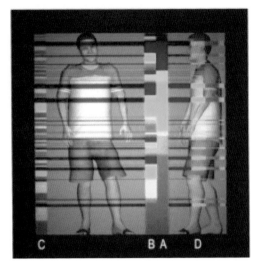

"This calls for wisdom. Let the person who has insight calculate the number of the beast, for it is the number of a man. That number is 666." (*Book of Revelation, 13:18*)

The next picture shows **chakra 7**.

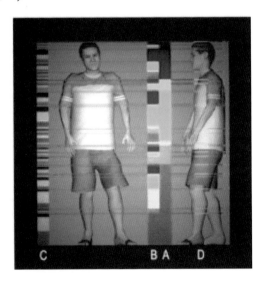

Here are a few more detailed pictures that give you an idea of the complexity of the human body. Where the chakras show themselves can be either a narrow or a wide band.

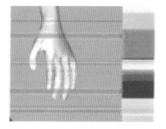 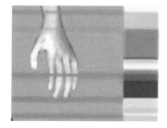

We can also clearly see that each of the seven chakras manifests in each of the seven colours of the spectrum. The composition per chakra is, relatively speaking, always the same, and yet the material expression in the various colours is different each time. Clever isn't it!

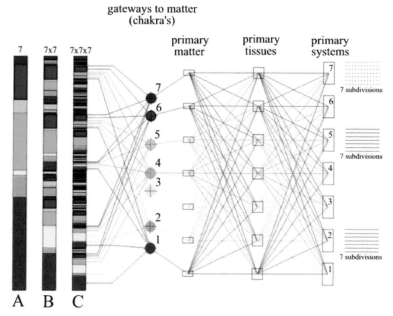

A – light spectrum
B – human field
(code 6B and 2D – 3 not developed yet – 7 in development)
C – individual human field

It is the constant pressure on the creative energy field that is responsible for the division of the $7 \times 7 \times 7$ areas, which are shown here as the human energy field represented in the correct code format

using 2D and 6B. We have shaded the colours for the frequencies
of 3 and 7 because these are not in development at all, as is the
case for 3 (the seed of the Universe – the seventh day), or only
partly developed, as in 7 (the human development). Therefore, B
represents the composition of the human field, which can then be
divided further into another seven parts for each layer. This gives
us C, where we view the human field in a much more detailed form.
The information concealed within the colours is then collected by
energetic pathways, which we called **chakras**. That is where the
energetic field gets compressed into matter, the PRIMARY MATTER
as we have seen. The first physical manifestations are the smallest
atoms, frequency 1, which are soon followed by connecting simple
atoms into simple molecules, such as water, oxygen and salts; frequency 4. Each layer introduces another complexity to the physical
structures as the layers of the field one by one manifest themselves
in a physical form, frequencies 1–4–6–2–5–7–3.

As a result of the increasing pressure, the physical manifestations are also encouraged to take up less and less space, which
results in them organising into groups, again ranging from simplest
to more complex, from lightest to heaviest. These are then called
the PRIMARY TISSUES, from which the PRIMARY SYSTEMS will be
formed later on, first of all in minute form within the cell – the
smallest space-occupying organisation that can be found – which
will then expand into more and more complex structures as the
pressure doesn't cease.

According to the division we have within the frequencies, we
can now understand the development of specific systems in the
macro-world. System 6, for instance, is divided into two parts,
and shows up in the later developments of animals and humans
as the digestive system (6A) and the respiratory system (6B). The
next layers within the development, 2 and 5, will show seven large
separate substructures because within 6 for the first time the
seven systems have been developed. Once developed, that part of
the development is then carried through in all subsequent layers.

Once frequency 7, the nervous system, has fully developed within the human being, seven clearly defined substructures will also be present. But that is for later.

First, I would like to briefly tell the myth of Osiris.

The myth of Osiris is the most well-known story in Ancient Egypt and tells about his death and resurrection; themes that represent the daily cycle of the death of the sun by sundown and its rebirth at dawn. Osiris was not only god of the kingdom and power of life to the pharaoh, he also personified the fertility of the land and the mind of the cycle of vegetation. As ruler of the kingdom of the dead he granted new life to those who earned immortality through the purity they displayed during their life.

Osiris was born out of Geb, the earth, and Nut, the goddess of heaven. As king of Egypt, Osiris was fair and truthful when creating the laws for his people. He knew fame and he knew jealousy. His work as king meant that he had to travel a lot. On returning from one of his journeys, Osiris was invited by his brother Set and 72 other conspirators, to a banquet. During the banquet, Set showed them all a beautiful, richly decorated coffin. After everybody had admired the coffin he promised to donate it to whomever it would fit. Everybody took turns to lay down in it but it fitted nobody. When it was Osiris' turn he stepped into it and stretched out. He fit into it perfectly. The conspirators then jumped onto the coffin, nailed down the lid, weighted it with lead and threw it in the Nile. When Isis heard what had happened to her husband she cut off a lock of hair and dressed in mourning.

After the death of Osiris, Set ruled over Egypt as a cruel tyrant. Isis fled and hid in the Nile delta. In the meantime, the coffin containing the body of Osiris ended up on the coast of Byblos, where a young cedar tree encircled it whilst it grew into a fine tall specimen of a tree. The king of Byblos admired the tree, cut it down and used the trunk to support the roof of his palace.

Gods and demons spread the news of this and so it came to Isis'

ears. In order for her, a goddess, to be able to enter the palace, she
had to devise a cunning plan. She went to Byblos and sat down by
a spring, where she spoke to no-one but looked after the female
servants of the queen. She plaited their hair and perfumed their
skin with the finest of scents. When the queen noticed how lovely
her servants looked she sent for Isis and employed her. She was
asked to take care of the queen's baby. At night Isis transformed
into a swallow and flew around the trunk that contained the
body of Osiris. However, the queen spied on her and discovered
who she really was. Afraid to have insulted a goddess, the queen
offered Isis to take whatever she desired from the kingdom. Isis
chose the large, beautifully engraved cedar pillar from the palace
and then she broke it in half. Inside, she found the coffin with the
body of Osiris. She placed the coffin on a boat and, taking with her
the eldest son of the king, she sailed away. At dawn a fierce wind
blew over the river Phaidros, which enraged Isis so much that she
steamed the riverbed dry. Arriving in a deserted area, she opened
the coffin and kissed her deceased husband. The son of the king
watched her carefully but she flashed daggers at him with her eyes
that killed him instantly. Isis hid the coffin with the body of Osiris
in the reeds of the River Nile.

Set was out hunting one night when, by the light of the moon,
he came upon the coffin. Enraged, he cut the body into 14 pieces
and spread them across the Nile delta. When Isis found the empty
coffin she did her utmost to gather the pieces back one by one. She
found them all except one – his penis. The symbol of his vitality
had been lost. She used all her magic powers and managed to
conceive from him a child, named Horus. After Horus was born,
Isis hid him in the neighbourhood of Buto, in the swamps of the
Nile delta, where he was guarded and protected by seven poison-
ous scorpions. When she returned to Osiris, Isis embalmed and
mummified her husband, which gave Osiris, protected by Ra as
the King of the Dead, eternal life.

When Horus became an adult he visited Set and demanded

the throne as his legitimate right. Set refused. After a long drawn-out battle, in which Horus lost one eye and Set lost his testicles, Horus defeated his uncle and succeeded his father, Osiris, as King of the Living.

Sources

Brunner-Traut, E., *Egyptian Tales.*

Foster, J.L. 2001, *Ancient Egyptian Literature: An Anthology*, pp. 103–109, Austin, Texas.

Lichteheim, M. 1976, *Ancient Egyptian Literature: A Book of Readings*, vol. II, Berkeley.

Naezer, J. 2006, Osiris, de mythe van, Den Haag.

Plutarchus, *De Iside Et Osiride.*

My interpretation of the story is that Osiris is the symbol of a flesh and blood person, in the Bible referred to as the Son. He was cut into 14 pieces by his brother Set, who is the incarnation of evil (the devil). When we look at the human body and we place it inside the visible spectrum, we notice that it can be divided into seven pieces twice (division A). We have seven divisions, the seven colours, on the left side of the body and seven divisions on the right.

By adding insight (Isis) to it, Osiris transforms into Horus. Horus, the reincarnation, then fights his uncle Set (evil). Horus loses one eye and Set loses his testicles. Horus has one eye remaining, which represents the third eye, which symbolises intuition and insight. The testicles relate to reproduction. Evil was stopped by insight. This means that we commit evil unwittingly, unaware of what we are doing. We could describe this transformation in one word, "osiritation". By giving the human being insight, we beat evil and rediscover our divine heritage of *God Created Mankind After His Own Image.*

Chapter 4
The Creation of Tissue

When we scan old writings for information on how things were created it always begins with energy. Ayurveda, the oldest written philosophy and healing system, describes in detail how life is structured, how it came about and what the most important elements are that constitute evolution.

Evolution is a manifestation of potential. Everything that exists holds within itself everything else. In a modern version we talk about the universe being a hologram. Just as we can create pictures whereby the overall picture is made up of smaller versions of the same picture, so does the universe hold a picture of itself in every detail. This is what the ancient texts from almost ten-thousand years ago tell us. Within the seed the tree already exists, and within the tree the woods are represented. The intelligence of the large is hidden within the small. One emerges from the other and has no other choice in the matter. This is the reality of creation; this is the power of creation. Life, creation and evolution are phases in the growing consciousness. Life is interdependent, completely connected, a two-way system of feeding and caring, sustaining each other. Everything is one and everything comes from that oneness.

71

This inherent intelligence is the consciousness of life itself. It is the pure feeling that already exists in plants and that is hiding within the rock, even within the atom. The attraction and repulsion of elements is comparable to love and hate, liking and disliking, gravity and radiation. It is for this reason that ancient seers of India believed that only the Self existed and that unity is the basis for all existence. Unity of life is the unity of consciousness, meaning that all is consciousness.

The human being as a microcosm contains within it all elements, all minerals, all plants and all animals. Within a plant there lies the potential to create a human being. And within the human being we can recognise the energy structure of the plant. Parts of the human body even resemble structures that are essential in the plant realm. For example, our circulatory and peripheral nervous systems are shaped like a tree, with a trunk that is split up into various branches. These systems in different structures have energetic connections that communicate directly with one another. Every bit of life is a radio station that receives and sends messages; feeding and supporting all. Everything exists because of the input of all other things, and the other way around too – everything is fed by every little living thing. In this way everything is interconnected in energy. Life is essentially light and cosmic energy. Earth itself is a receiving and sending station; it breathes cosmic forces in and out, which allows life forms to grow within this breath.

The creation is light. In the Vedas, the old teachings from India, the great god Agni, the principle of fire, creates the world and manifests the whole of creation through a series of transformations of the self. Plants transform light into life. Human beings transform life into knowledge. Here we have the three fundamental forces – light, life and love – whereby each is an extension of the other; three dimensions of the same existence. Christianity tells us about the Holy Trinity of the Father, Son and Holy Ghost. The Father represents the principle of life (energy). The Son represents the principle of love (togetherness), and the Holy Ghost represents the principle of light.

The human being is the "knowledge" layer within creation. The plants work on a different level but in the same direction, representing consciousness. That is how they can feed our mind and tissues to develop this consciousness. *So Above, so Below*, the whole universe is a metamorphosis of light. In the outside world the sun is the source of light and life. In the inner world there also exists a sun as the source of life: it is our true self. Plants connect us to the energy of the sun, while our internal plant structure – our nervous system – connects us to our inner sun. This allows a free flow of consciousness, in and out, that liberates the mind and creates ecstasy in life.

The wise men of old India approached healing and herbs with that same consciousness. Their science was not built on experimenting, but on a knowledge of interconnectedness. Experiments indicate a distance, a separation between the observer and the observed, and this leads to methods, to measurements. Direct observation or meditation, however, is the science that allows things to show up as they really are.

The Creation of Light

Light was created before anything else in our universe, which is confirmed in our culture too. This has been made clear in the Bible as well as in other ancient texts that describe *the beginning of time*.

Light is part of an electromagnetic energy that engulfs the cosmos. We are only aware of a very small part of the entire spectrum. There is a small band of radio waves we can hear and a very small part, right in the middle of the spectrum, that we can see. Our senses only pick up a small range within the electromagnetic spectrum. For the great majority of cosmic waves, we do not have conscious sensors. However, this does not mean that these waves do not affect us or that they are inactive as far as we are concerned.

For instance, we know that ultraviolet light consists of three wave bands: A, B and C. Although the human eye is not capable of picking up these rays, we are well aware of the effects these have on our systems. Waveband A produces redness of the skin and it can make certain materials become fluorescent, whilst in other materials a photochemical reaction takes place as a result. Waveband B is responsible for the redness as well as the pigmentation of the skin, and stimulates the production of vitamin D, which is essential for the strength of the tissues in general. Waveband C has a germ-destroying influence.

By recognising that the radio waves we hear, the light we see and the ultraviolet waves we know of all influence our functioning, our mood and our life, we can open ourselves up to the fact that all other radiation also has a deep impact on us. These include, for example, the radio waves we don't notice, such as radar, micro-waves and all wavebands used for communication. Furthermore, there are infrared waves, photographic waves, x-rays, gamma rays and many different cosmic waves.

In many cultures, light has a connection with the manifestation of the divine, and the knowledge that will lead to enlightenment; a state of wholeness. Light has been regarded as the source of life itself. *Enlightenment* then becomes synonymous with health in terms of the body as well as the mind and the spirit. The sun, and the gods that relate to the sun, have been worshipped throughout the ages. Not only did people from ancient cultures organise special ceremonies dedicated to their sun gods, but they also built their houses and temples according to the influence of the sun. They lived their lives according to the seasons and the rhythm of day and night.

In ancient Greece, colour and sound were used together. The followers of Pythagoras, for example, worked with the knowledge of figures and created scientific theories about sound and musical octaves, which they used in their healing methods together with colour. Colour and sound are closely linked. Both are easily recognisable vibrational energies and every colour with its specific timbre and variations can be linked to a specific sound.

The use of colour treatment with sunlight (heliotherapy) was common in Greek and Roman times, and they are believed to have been the first to put down on paper the theory and the practical use of sunlight in healing. Herodotus has been said to be the father of this treatment. The Egyptian city of Heliopolis ("city of the sun") was famous for its temples that were designed to break down sunlight into the colours of the spectrum, just as the Atlanteans and Egyptians did before them. During the first five centuries of Christianity, all medical use of colour, chanting and worshipping various gods was seen as heresy. Consequently, a lot of Greek and Roman scriptures on holistic medical therapies were destroyed.

During the middle of the nineteenth century, treatment with sunlight was reintroduced by Jacob Lorber in his book *The Healing Force of Sunlight*, which was first published in 1851 in Germany. The "father of photobiology", Danish doctor Niels Ryberg Finsen, was the first to develop light treatment in a scientific way. In 1892 he began treating skin tuberculosis using artificial light produced by carbon.

Traditionally, people spent most of their time outside, where the colours of nature dominated life. The changing seasons also meant that the influences of weather, sunlight, rain and snow, wind and temperature, were all reflected in the colours of their environment. We notice the bright colours of summer, orange and brown in the autumn, grey and dark brown in winter and the green and yellow in spring. Why does it happen like that? How can we begin to understand the harmony we are surrounded by?

The visible light spectrum is divided into seven colours, each with its own frequency. At the bottom end of the spectrum, with the longest wavelength and the lowest frequency is red. When we move up the frequency ladder, we change from red to orange, to

yellow, to green, to blue, to indigo, to violet. An interesting observation is that the width of the wave bands can be calculated with
the Golden Ratio, as explained in the previous chapter. This leads
to an increasing width of the band, starting with the smallest one,
being yellow, moving up to violet, to blue, to orange, to indigo, and
to the largest part, which is red.

It is an important fact that the Golden Ratio, which is used in
mathematics, physics and art, also plays a significant role in nature.
Adolf Zeising proved that the Golden Ratio is how branches are
divided along the stem of the plant, and how the nerves of a leaf are
divided. He found similar relationships in the skeletons of animals
and in the divisions of their blood vessels and peripheral nervous
systems, but also in the properties of chemicals and in the geometry
of crystals. He defined the Golden Ratio as a universal law. In 1834
he wrote: "*The Golden Ratio Is a Universal Law That Contains the
Basic Principle of Everything That Is Formed and It Strives Towards
Beauty and Completion in Nature as Well as in Art, and That Like a
Superior Spiritual Ideal Penetrate All Structures, All Forms and Properties, Irrespective If They Are Cosmic or Individual, Organic or Inorganic,
Acoustic or Optic, but its Ultimate Shape Is the Human Form.*"

The visible light spectrum is created from the small yellow
band in the middle. This is the fertile part of creation. It is, so to
speak, the genital part of creation. In the human being we would
recognise it as the seed-producing organ from which all other
tissues are created. From this the whole cycle of development
starts. The yellow colour is therefore the equivalent of the tissue
of sexual fluids. What other tissues are required to complete the
human form? According to Ayurveda, all creation cycles have seven
tissues, whereby one is formed out of the previously existing one.
This happens in plants, animals and humans.

From the Golden Ratio we know that colours emerge in a fixed
order, one after the other, starting with yellow. Next is red, then
green, indigo, orange, blue and violet. These appear when light
travels through a denser medium, such as a water droplet. It is the

density of the medium that brings out the information that was previously hidden within it. The cosmos is a very dark place. In air, the atmosphere around the earth, light becomes visible as white light. Using a denser medium, such as water, the various frequencies from the white light are filtered out. These become visible as individual bands of colour. Going into even denser material, such as earth, light cannot penetrate any further.

Each of the colours relate to a layer of tissue, which is created from the energies contained within the spectrum of light; the creator of things within this universe. When the energies move into higher-density areas, the tissues are "filtered" out. It is on this basis that we see the colour red as being responsible for the first tissues, which are juices. GREEN follows and creates blood tissues; indigo creates muscle tissues; orange creates fat tissues; blue creates bone tissues; and bone marrow and nervous tissues are created out of violet. As mentioned previously, the sexual fluids relate to the colour yellow.

Now, every colour is made up of all the other colours, only in different proportions, which results in either white light or the individual colours. This means that green will contain mostly green and less of the other colours. The main colour you see will always have the highest contribution of that colour.

Ancient knowledge has shown us that all living things have energy centres from where the tissues are directed. Energy centres that we relate to the body are known as *Chakras*. Each of these are related to a specific sound and a specific colour. There are seven main chakras for the human being, as explained previously. These chakras deliver energy to the system and make it function. So, seven chakras, seven colours and also seven tissues. Chakras occupy specific places in the body between the crown (top of the head) and the base (bottom of the pelvis) and relate to crossover points of important energy lines. Traditionally, every chakra has its own colour and these appear as the colours of the rainbow, from bottom to top: red, orange, yellow, green, blue, indigo and violet.

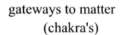

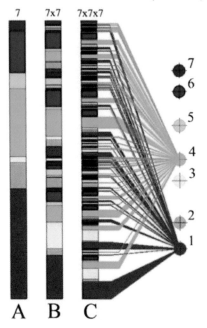

A – light spectrum
B – human field (every colour divided up again)
C – development of the human field (every colour divided up again!) Energies are being compressed into chakras (energy centres).

In accordance with the universal laws of creation, we can now conclude that the order in which these chakras manifest must be the same order as the colours of the rainbow and the tissues come into being. From yellow (3) appears first red (chakra 1), then green (chakra 4), then indigo (chakra 6), followed by orange (chakra 2), then blue (chakra 5) and finally violet (chakra 7), before yellow appears again (chakra 3). Each chakra hands down information and also filters information, which brings out the various aspects that will eventually result in the formation of tissues. Traditionally, the chakras are known to possess and deal with specific energy information, and when we look at what that represents, we can see how in the creation lower density evolves towards higher density, always following the same sequence from 1 down to 7 and 3.

- The first chakra in creation (1) relates to *Grounding and Surviving* (group identity, emotional bonding, loyalty).
- The next one (4) relates to *Balance* (trust, forgiveness, hope, compassion).
- Next comes one that relates to the *Consciousness* (6) (inspiration, fantasy, wisdom).
- Next is a chakra that relates to *Personal Growth* (2) (desire, perseverance).
- Then we come to the chakra relating to *Communication* (5) (expression of self, self-knowledge, expression of willpower).
- And then it is the chakra relating to *Knowledge and Dedication* (7).
- We end up with the last chakra (3) representing *Individual Power and Feeling* (self-worth, self-discipline, courage).

Note how the energies condense from group towards individual, and from principles in life towards real physicality.

This gives you an idea of the path that light follows to create everything within the universe, from the cosmic dance of the stars to the amoeba. Dividing of visible light is but the first step in a long creative process within the universe. Again and again it repeats itself, according to the same principles of increasing density. This creates layers of sevens, from light to chakras to tissues.

We live in a light universe. Light serves as the boundary of our universe. Think about it: nothing can be without light, and nothing travels faster than light in this universe. Everything within the boundaries of this universe is determined by light.

Creating Plants

Nature consists of three basic qualities, or three prime attributes. First, there is the principle of light, perception, intelligence and harmony. Second, there is the principle of energy, activity, emotion and turbulence. Third, we have the principle of inertia, darkness, dullness and resistance. These principles can be recognised within individuals as well as in plants or any other parts of the universe. Those people for whom the principle of light dominates, give value to truth, honesty, humility and the good of all. Those who have a predominance of the principle of energy, value power, prestige, authority and control. Those dominated by the principle of inertia remain trapped in fear, servility, ignorance and the focus of decay.

From these three principles arise five elements. From light, consisting of clarity, comes the element of **Ether**. From energy comes **Fire**. From inertia comes **Earth**. Between light and energy arises the subtle but mobile element of **Air**. Between energy and inertia arises **Water**, combining mobility with inertia. These elements are the five states of matter: solid (earth), liquid (water), radiant (fire), gaseous (air) and ethereal (ether). They delineate the five densities of all substances, all visible and invisible matter in the universe. The five parts of plants show how their structure relates to these five elements: the root corresponds to earth, the densest part of the plant; the stem and branches are the water element, as they allow the sap to move; the flowers correspond to the fire element, which manifests light and colour; the leaves are the air element, since the breath of the plant moves through them; the plant breathes through the leaves; and the seed corresponds to the ether element as it is the subtle essence of the plant.

As the elements can be related to the structure of a plant, the elementary tissues of a plant are also formed in a specific order and format. This is no different from the way all matter is formed in the entire universe. There are seven concentric layers of tissues, created from the outside in. Each consecutive layer arises out of the previous one and is denser than its predecessor. As the energy condenses, it creates subsequent layers of tissues; seven in total: the most subtle on the outside and the densest on the inside, right at the core of the energy field.

The reproductive tissue is the essence of all bodily tissues and contains within itself not only the power of reproduction, but also that of rejuvenation. From the reproductive tissues originate all other tissues, thereby creating the world as we observe it.

The outer layer of tissues is formed by the juice of the plant, its plasma. The second layer is the resin of the plant, its blood. The third is the softwood, its muscle. The fourth is the gum, its fat. The fifth is the bark, its bone. The sixth are the leaves, its marrow and nervous tissue. And the seventh layer is the flowers and fruits, its reproductive tissue. From this layer a new series of seven layers will be formed to create another similar plant in the same way.

Within the human body, we find these tissue layers are also well defined. Dr Robert Svoboda describes the tissues and their functions as follows:

Juices – sap or juices; tissue fluids, including chyle, lymph and blood plasma. Its accessory tissues are breast milk and menstrual blood, and its waste product is mucus. Its function is nourishment.

Blood – red blood cells. Its accessory tissues are blood vessels and tendons, and its waste is bile. Its function is invigoration.

Flesh – skeletal muscle. Its accessory tissues are ligaments and skin, and its wastes are those that accumulate in body orifices, such as ear wax, snot, navel lint, smegma and so on. Its function is "plastering" of the skeleton.

Fat – fat in limbs and torso. Its accessory tissue is omentum, and its waste is sweat. Its function is lubrication.

Bone – all bones. Its accessories are the teeth, and its wastes are body hair, beard and nails. Its function is body support.

Marrow – anything inside a bone: red and yellow bone marrow, and the brain and spinal cord, which are wholly encased in bone. Its accessory tissue is head hair, and lachrymal secretions are its wastes. It performs "filling" of the bones.

Shukra – male and female sexual fluids. Its tasks are reproduction and production of the subtle essence of all vital fluids. It has neither accessory tissues nor waste products.

Ayurveda states that each tissue is formed from the one immediately previous to it, except the accessory tissues, which are only nourished and do not nourish in return. Breast milk and endometrium are meant for nourishing a child, not for the nutrition of any part of the mother. Excess of any waste at each stage is indicative of poor digestion at the level of that tissue.

There are three ways in which one tissue layer can feed the next. A layer may completely convert itself into another. For instance, the juices provide direct nourishment for the production of blood. A large part of circulating juice is converted into blood. Or the tissue may flow through the body, gradually nourishing the next tissue in line through a more complicated series of reactions. When blood nourishes flesh it flows through many different parts of the body, nourishing all flesh along the way. Or the tissue may simply "seed" the next tissue by sending hormonal or enzymatic cues to it. This is how flesh nourishes fat and fat nourishes bone.

Creating Everything

Everything in our universe is energy and all matter is created out of this energy. By increasing the pressure and/or lowering the temperature, energy is condensed and eventually some of it might become matter. This has to do with waves and frequencies as we experience them within the electromagnetic spectrum of our universe. In order to understand frequency, it is first necessary to comprehend two related varieties of movement: oscillation and wave motion. Both are examples of a broader category – periodic motion: movement that is repeated at regular intervals called periods. In chaos theory, scientists have shown that all movements, no matter how complex they are, are part of a repetitive process, that it is now believed that *All Movement Is Part of a Cycle* that repeats itself, in the same way that a "straight line" is a small part of a large circle.

Even though many examples of periodic and harmonic motion are found in daily life, the terms themselves are certainly not part of everyday experience. On the other hand, everyone knows what "vibration" means: to move back and forth around a place of stability. Oscillation is simply a more scientific term for vibration. While waves are not themselves merely vibrations, they involve, and may produce, vibrations. This is, in fact, how our sensory organs work: by interpreting vibrations that result from waves.

Indeed, the entire world is in a state of vibration, although people seldom perceive this movement, except perhaps in dramatic situations such as earthquakes. All matter vibrates at the molecular level, and every object possesses what is called a natural frequency, which depends on its size, shape and composition.

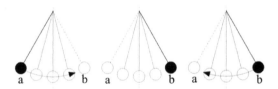

Oscillation is a type of harmonic motion. Traditionally, it is thought of as two basic types: that of a swing or pendulum and that of a spring. In both cases an object is disturbed from a position of stable equilibrium and, as a result, it continues to move back and forth around that stable equilibrium position.

But, as we have already seen, a straight line is part of a large circle and, therefore, these two types can be seen as one and the same. All oscillation is a pendulum movement, but sometimes the anchor point is so far removed from the observed movement that it appears as if the movement is in a straight line (up and down). In oscillation, there is always a cycle in which the oscillating particle moves from a certain point in a certain direction, then reverses direction and returns to the original point. Usually, a cycle is viewed as the movement from a position of stable equilibrium to one of maximum displacement, or the furthest possible point away from the stable equilibrium and back again. Because stable equilibrium is directly in the middle of the movement path, there are two points of maximum displacement and within one cycle they both have to be reached.

Wave motion is a type of harmonic motion that carries energy from one place to another without actually moving any matter. While oscillation involves the movement of "an object", a wave may or may not involve matter. An example of a wave made out of matter, a mechanical wave, is a wave on the ocean, or a sound wave, in which energy vibrates through a medium such as water or air. Even in the case of a mechanical wave, however, the matter does not experience any net displacement from its original position. Water molecules end up where they began.

Periodic waves may be further divided into transverse and longitudinal waves. A transverse wave is the shape that most people imagine when they think of waves: a regular up-and-down pattern (called sinusoidal) in which the vibration or motion is perpendicular to the direction the wave is moving. A longitudinal wave is one in which the movement of vibration is in the same direction as the wave itself. Though these are a little harder to picture, longitudinal waves can be visualised as a series of concentric circles emanating from a single point. Sound waves are longitudinal. This is an artificial distinction created by the point of observation. When we look along the direction in which the wave is travelling in a medium, we have the impression that the oscillation is in the direction of the wave. However, the matter is only being pushed up and down, in a direction perpendicular to the wave movement, just the same as with a sinusoidal wave, but here we look at the wave movement sideways. The wave itself moves away from the observation point.

Though waves and oscillators share the properties of amplitude, period and frequency, the definitions of these differ slightly depending on whether we are discussing wave motion or oscillation. Amplitude, generally speaking, is the value of maximum displacement from an average value or position. Amplitude is "size". For an object experiencing oscillation, it is the value of the object's maximum displacement from a position of stable equilibrium during a single period. For a wave, it is the maximum value of the pressure change between waves.

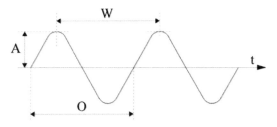

A: amplitude; W: wavelength; O: oscillation

Other than with amplitude, period has a direct relation with frequency. For a wave, the period is the time interval between waves. With an oscillator, the period is the amount of time needed to complete one full cycle. The value of a period is usually expressed in seconds.

Frequency in oscillation is the number of cycles per second, and in wave motion it is the number of waves that pass through a given point per second. These cycles per second are called Hertz (Hz) in honour of the nineteenth-century German physicist Heinrich Hertz.

When waves travel from one medium to another, their frequency remains exactly the same, but the wavelength and the speed change. Thus the *Manifestation* of the frequency is different. This is why increased pressure on energy results in a condensation of energy and eventually in matter with a shorter wavelength and a higher speed, but the same frequency. Matter, therefore, is specifically related to the energy it comes from.

When We Increase the Temperature of the Environment, a molecule will vibrate with a longer wavelength at a slower speed, thereby keeping the frequency the same. Once again, the matter will become "lighter", but it will carry the same energy information, only in a different form, meaning it is still the same matter. If, however, *We Increase the Temperature of a Molecule*, of the matter itself (adding energy to it by altering the electromagnetic constitution of the molecule), then the molecule will vibrate at a faster rate, at a higher frequency, through an unaltered medium, thereby changing the energy information it is carrying. This means the molecule has changed into a different one.

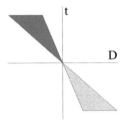

There is an abrupt change from one frequency band to the next. First, there is instability in energy or mass when moving away from the balance point. Once it clears the transformation hole it changes expression.

There are two ways in which the frequency can be changed. We know from music that when we double or halve the frequency we meet the same note but an octave higher or lower. The same note means that the new frequency carries the same basic information but either in a "looser" or "more condensed" way. In between the first frequency and the same frequency an octave lower, different changes can be made, each step resulting in a different note. There are twelve equal steps to move down an octave, resulting in seven major different notes and five semi-notes. Musical frequencies do not appear as a continuum of frequencies, one smoothly melting into the next; they are completely separate frequencies (notes).

The quantum theory, created by Erwin Schrödinger, Werner Heisenberg and others, reduced the mystery of matter to a few postulates. One is that subatomic particles have both particle- and wave-like qualities, obeying a well-defined equation, which describes the probability that certain events occur. Another is that energy is not continuous, but occurs in discrete bundles, called "quants". (The photon, for example, is a quantum or packet of light.) Electrons can only orbit the central nucleus at certain paths. To change the orbit of an electron to a higher path, a fixed amount of energy will need to be added to the atom. Similarly, an electron that "drops" an orbit releases a specific amount of energy. This is the quantum.

Thus, quantum theory confirms the relationship between the various notes in as much that each note is separate from the previous one and differs from it by one "step", one packet of energy, which, in music, can be heard as a note of a slightly different frequency.

And the other way around, quantum physics now knows that there are only twelve different steps in any cycle, before we return to the beginning of the cycle once again. From these twelve steps only seven will manifest in the physical world; the other five exist but have a supportive role. We need to explore our world, our universe, in the order of seven and twelve. Only then will we be able to understand the structure, the functioning, and we will be able to predict evolution.

The Origin of Species

Creation is such a massive concept that the small human mind and the even smaller human consciousness is incapable of completing an overview of it, let alone of classifying all that exists within creation. It not only appears to be an impossible task to know and classify all plants, it is even less possible to have an insight into the organisation of the animal kingdom. We cannot even pretend to know all species, so how can we then be sure to number them all and name them correctly? In order to make definite progress in that direction we need to be able to catalogue what we observe in a "logical" manner, which could easily include all we don't even know.

This, of course, starts with a proper understanding of creation itself: what came first and what belongs with what, which characteristics unite and which separate. Darwin developed his theory of creation on this principle and came up with a story about the origin of the species. From a great number of studies since that time, we now know that there are serious gaps in his explanation of how the various species have been able to develop. Some of the intermediates that are suggested by the theory are sadly lacking, or have never been found. Now that we have suggested a path of creation it would be logical that we also look at how that would lead to classification of plants and the animal kingdom.

We have come to a code for the creation. This has emerged from the materialisation of colours within the visible light spectrum. We can derive from that what kind of energy is being expressed on what day of the creation, and what the energetic composition is of each of those days or layers.

For instance, we know that the creation code can be written as 1–4–6–2–5–7–3, coming forth from 3' energy, the seed of the universe, which gives rise to the various material levels of the creation:

Day 1	energy layer 1	colour: red	form	heaven and earth
Day 2	energy layer 4	colour: green	balance	water and land
Day 3	energy layer 6	colour: indigo	consciousness	plants
Day 4	energy layer 2	colour: orange	movement	egg-laying animals
Day 5	energy layer 5	colour: blue	communication	mammals
Day 6	energy layer 7	colour: violet	knowledge	human beings
Day 7	energy layer 3	colour; yellow	personal power	

Via our studies, we have been able to actually place the seven energy centres, from which matter manifests, on very specific positions in relation to each other. We used the well-documented chakra positions for this purpose and calculated the exact physical positions between crown and base. This led to an insight into how the composition of each of these energy layers becomes evident, and we can visualise this by placing, in order of appearance, the numbers as they occur.

By drawing concentric circles with each of these energy centres at their centre, we can visualise how the energy waves that leave such a centre spread through space and which other centre they reach first, second, third, etc. By using the Golden Ratio, we can calculate the contribution percentage for each energy band  to each centre, because the highest contribution will be 39.55%, followed by 24.45%, then 15.11%, then 9.34%, then 5.77%, then 3.57% and finally 2.20%. Now, we can fill in the contribution that each centre has in the functioning of each of the others. From point 1, we first reach point 3, then 4, 6, 2, 5 and 7.

The 1st Day: Heaven and Earth

The code for this energy layer 1 is 1–3–4–6–2–5–7. This represents the creation of heaven and earth, energy and matter. This is a first expression of any physical manifestation, which are the atoms and the gases. This is the beginning of all material creation.

To arrive at a full energy layer 1, it requires seven phases all together to complete day 1. Let us have a look at those steps.

The first phase 1.1 can be described as developing through steps 1.1.1, 1.1.3, 1.1.4, 1.1.6, 1.1.2, 1.1.5, and 1.1.7. This is 1.1 completed (phase 1 of day 1).

The following drawing shows the subsequent steps and the activated frequencies. In the upper drawing, the arrow on the left indicates the sequence of the different steps, starting with 1.1.1, which activates an area with the red band, followed by 1.1.3, which activates an area within the yellow band, and so forth.

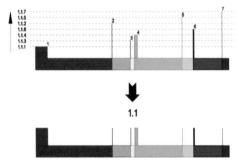

1.1

The second phase 1.3 can be described as using steps 1.3.3, 1.3.1, 1.3.4, 1.3.6, 1.3.2, 1.3.5, and 1.3.7. This is 1.3 completed (phase 2 of day 1).

The next drawing shows the subsequent steps and the activated frequencies:

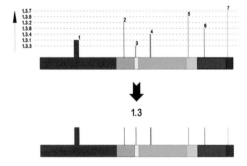

1.3

The third phase 1.4 can be described as using steps 1.4.4, 1.4.6, 1.4.1, 1.4.2, 1.4.3, 1.4.5, and 1.4.7. This is 1.4 completed (phase 3 of day 1).

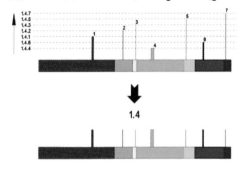

The fourth phase 1.6 can be described as using steps 1.6.6, 1.6.4, 1.6.2, 1.6.1, 1.6.5, 1.6.3, and 1.6.7. This is 1.6 completed (phase 4 of day 6).

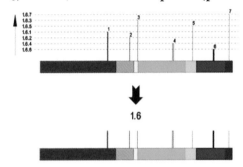

The fifth phase 1.2 can be described as using steps 1.2.2, 1.2.6, 1.2.5, 1.2.4, 1.2.7, 1.2.1, and 1.2.3. This is 1.2 completed (phase 5 of day 1).

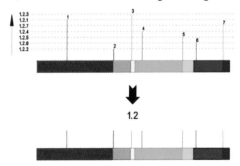

The sixth phase 1.5 can be described as using steps 1.5.5, 1.5.7, 1.5.2, 1.5.6, 1.5.4, 1.5.1, and 1.5.3. This is 1.5 completed (phase 6 of day 1).

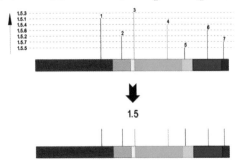

The seventh phase 1.7 can be described as using steps 1.7.7, 1.7.5, 1.7.2, 1.7.6, 1.7.4, 1.7.1, and 1.7.3. This is 1.7 completed (phase 7 of day 1).

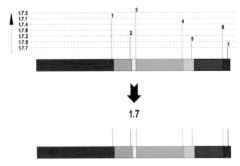

The first day is completed in seven phases: 1.1, 1.3, 1.4, 1.6, 1.2, 1.5 and 1.7.

The next drawing shows the sequence of these phases together with their activated areas (frequencies) within the visible light spectrum:

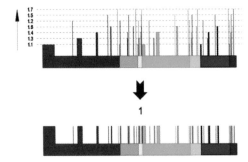

The 2nd Day: Water and Land

The code for this energy layer 4 is 4–6–1–2–3–5–7. This represents the creation of water and land. The material part of layer 1 will be further divided and developed into water and sand. Air or gases (frequency 1) remain manifest as the rest becomes, in the first instance, fluid from which all other matter will emerge.

The second day can be described as using the following steps 4.4, 4.6, 4.1, 4.2, 4.3, 4.5 and 4.7. These are the seven phases of day 2 and every phase consists of seven sequential steps.

- Phase 1 (4.4) contains the steps 4.4.4, 4.4.6, 4.4.1, 4.4.2, 4.4.3, 4.4.5 and 4.4.7.
- Phase 2 (4.6) contains the steps 4.6.6, 4.6.4, 4.6.2, 4.6.1, 4.6.5, 4.6.3 and 4.6.7
- Phase 3 (4.1) contains the steps 4.1.1, 4.1.3, 4.1.4, 4.1.6, 4.1.2, 4.1.5 and 4.1.7.
- Phase 4 (4.2) contains the steps 4.2.2, 4.2.6, 4.2.5, 4.2.4, 4.2.7, 4.2.1 and 4.2.3.
- Phase 5 (4.3) contains the steps 4.3.3, 4.3.1, 4.3.4, 4.3.6, 4.3.2, 4.3.5 and 4.3.7.
- Phase 6 (4.5) contains the steps 4.5.5, 4.5.7, 4.5.2, 4.5.6, 4.5.4, 4.5.1 and 4.5.3.
- Phase 7 (4.7) contains the steps 4.7.7, 4.7.5, 4.7.2, 4.7.6, 4.7.4, 4.7.1 and 4.7.3.

Take note of the relationship between the code and the place in the spectrum of the visible light where the bands of frequencies

manifest themselves. Take, for instance, phase 1 (4.4) where the
first frequency band (4.4.4) manifests in the green part of the
spectrum, the second one (4.4.6) in the indigo part, the third one
(4.4.1) in the red part, and so on.

The various phases of day 2 are shown in the next upper drawing
and the activated areas (frequencies) of the visible spectrum in
the lower one.

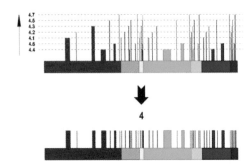

The 3rd Day: Plants

The code for this energy layer 6 leaves us with two equal possibilities:

6–4–2–1–5–3–7, which we call **6A**; and
6–4–2–1–5–7–3, which we call **6B**.
Six represents plant life.

Here, the fluid part becomes denser, or more compact. Matter
becomes fixed and all that we experience as "concrete matter"
begins its development here. This starts with single-celled organ-

isms such as bacteria, which develop step by step into plants. Plants develop in water as on land, which is in fact in air. In their development they quickly attach their roots in earth, whilst the plant itself grows in either water or air. Some plants drift in water and some have drifting roots, but in air there are only a few species with drifting air roots. Creating this plant life is possible following one of two routes: 6A and 6B. What effect would this have on the developing plants? The difference relates to the last two digits in the code; the least-contributing energies to the plant structure. This means that both ways actually result in two manifestations that are pretty similar in appearance, but differ in details. What kind of details? We are looking at the input of energies 7 and 3. Seven relates to knowledge and three to personal power. Frequency 7 manifests as the nervous system and 3 as the seed. The power of reproduction lies within frequency 3, but only when 3 is also the smallest part of the whole. In other words, it needs to be placed in the last position of the code for it to be able to manifest reproduction.

The first two layers of creation, as discussed above, end in 7, which means that these layers cannot reproduce themselves. The *Universe* can create new water, but water itself cannot reproduce. In code 6, in plants, half the material expression is incapable of reproduction, but the other half, ending in 3, is. In plants, the seed is represented in the flowers and fruits, which are present on the top of the plant. The plant above the soil and the roots are very similarly structured when we compare the shape of the branches to the extensions of the roots in the soil. However, the roots cannot generate flowers or fruits. They cannot reproduce the plant. This part of the plant code ends in 7. The plant develops a top part and a distinct bottom part. From here on, the development of all other creatures will have a distinct top and bottom part – heads and tails – because frequency will manifest in every other layer of the creation and thus in every other species that will develop later. The amount of contribution frequency 6 delivers will vary throughout the various layers of development.

The third day can be described as 6.6, 6.4, 6.2, 6.1, 6.5, 6.3 and 6.7, but also as 6.6, 6.4, 6.2, 6.5, 6.7 and 6.3.

The following drawings show us the various subsequent phases of day 3 in both possibilities of 6A and 6B:

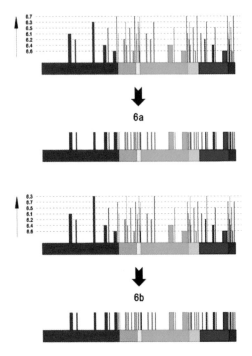

In the next drawing we can see where the differences between 6A and 6B manifest within the spectrum:

Next is a detailed drawing of the area of difference in manifestation between 6A and 6B. We notice that for 1 and 2 there are no differences, but 3, 4, 5, 6 and 7 all shift and that there is a difference in width in frequencies 3 and 7, with 3 being larger in 6A and 7 being larger in 6B.

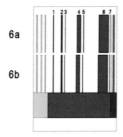

The 4th Day: Egg-Laying Animals

The code for this energy layer 2 gives us four possible pathways:

2–6–5–4–7–1–3, which we call **2A**;
2–6–5–7–4–1–3, which we call **2B**;
2–5–6–4–7–1–3, which we call **2C**; and
2–5–6–7–4–1–3, which we call **2D**.

The energies of 5 and 6 can be exchanged, as can the energies of 4 and 7. This layer of creation represents the emergence of egg-laying animals such as fish, birds, reptiles and amphibians, but also insects, worms, shellfish, jellyfish, etc. These animals occur in water as well as land (living in air) and emerge from the previous layer of development in both available environments: soil–water and soil–air.

How can we even begin to classify this great variety of animals so that we are sure to include all known but also all unknown species and give them a place in evolution? We need to consider the various energetic pathways whereby there is a great difference in the expressions of the positions of 5 and 6 as they occupy position two or three in the code. Frequency 5 represents communication and 6 consciousness, which in evolutionary terms creates plants. In terms of tissue production, 5 creates bone tissue and 6 muscle tissue. In 2C and 2D, frequency 5 is more prominently present than 6, which means that the species have a more-prominent bone structure. These we know as the vertebrates (they have a backbone), whereas 2A and 2B represent the invertebrates (those without a backbone).

The second division has a lesser impact on the physical manifestation because the exchange between 4 and 7 occurs much further down in the code. Seven represents knowledge and 4 is balance. In terms of tissues, 7 delivers the nervous system and 4 the circulation, the blood tissue. When we examine the nervous systems of the egg-laying animals, we can divide them into two clearly separate groups. One has a simple nervous system comprising ganglia, nodes of nervous tissue, and strings that connect places, usually one long string along the back of the animal and one along the front. Some animals have more complicated nervous systems whereby in one place a separation occurs of several ganglia, usually in the head. This is an early stage of a central nervous system, a brain, which structurally is totally different from the peripheral nervous system with its long strings of nervous tissue. The peripheral nervous system is then located outside of the central nervous system. Hence, in both categories, vertebrates (5) and invertebrates (6), we discover animals with a primitive nervous system (4 before 7) and others with a central nervous system (7 before 4). As everything is continually in evolution, we can conclude that all animals must fall into one of these categories:

- invertebrates with a primitive nervous system (2A);
- invertebrates with a central nervous system (2B);
- vertebrates with a primitive nervous system (2C);
- vertebrates with a central nervous system (2D).

It is obvious that most invertebrates we know of have a primitive nervous system (2A) and most vertebrates have a central nervous system (2D). In energetic terms, this means that when pathway 6 is chosen over 5, we most likely also choose 4 over 7. When we have a simple physical structure, it creates more diversity having a simple nervous system, as a more complex nervous system would quickly become difficult to be carried by the simple physical structure, as evolution has shown us. Similarly, choosing 5 over 6 encourages the choice of 7 over 4 because with a physically more complex structure it is easier to develop a more complex nervous system. One development concentrates on simplicity and the other on complexity. The interesting bit is that the other development lines are also present, although most species have become extinct because the combination of either a complex nervous system in a simple physical structure is difficult to sustain, as it is with a simple nervous system within a complex physical structure. Some of these species are, however, still around.

The fourth day can be described as:

    2A and formed by steps 2.2, 2.6, 2.5, 2.4, 2.7, 2.1 and 2.3.

    2B and formed by steps 2.2, 2.6, 2.5, 2.7, 2.4, 2.1 and 2.3.

    2C and formed by steps 2.2, 2.5, 2.6, 2.4, 2.7, 2.1 and 2.3.

    2D and formed by steps 2.2, 2.5, 2.6, 2.7, 2.4, 2.1 and 2.3.

In a drawing it looks like this:

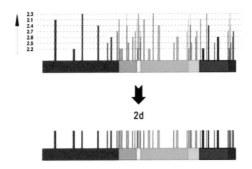

The 5th Day: Mammals

The code for this energy layer 5 is 5–7–2–6–4–1–3. This represents
the creation of the mammal.

Once frequency 2 has expressed those four possible ways, rem-
nants of that development can then be found within the subsequent
layers of evolutionary development, as is the case with mammals
and humans. So the first division was about 6 and 5, muscle and
bone. Looking at the structure of mammals and humans, we notice
that the back side is much more compact in structure than the
front. There is more bone in the back structure than there is in the
front; amongst other things there is the spinal column in the back.
This makes the front of the body an expression of energy 6, and
the back an expression of energy 5. Six represents consciousness
and also that our life is directed forward. Communication (5) is,
in mammals, more obviously done by the shape of the back. Think
about the hairs along the spinal column, about arching the back
ready to attack, about the way they hold their heads to warn of
potential danger.

We humans talk about "body language" or the way we express
ourselves in the shape we give our physical appearance. At the same
time, the differences between 4 and 7 can be found in our physical
structure too. Frequency 7 brings forth the nervous tissue and 4
the blood tissue, circulation. The heart is placed on the left side of
the body as an expression of 4, which would leave the right side to
be an expression of 7. We also know that 7 represents knowledge
and 4 balance, and ancient texts remind us that the left side of the

body represents the teacher and the right side the student. These texts too indicated that there is an energetic difference between the two sides of the body. Wisdom, in the East, is balance and thus frequency 4 is expressed through the left side of the body. Gathering knowledge and learning to become wise is an expression of energy 7, represented on the right side of the body.

Remember that differences from previous developments are expressed in this new layer too. This means that the various possibilities in 6 and 2 will find an expression within the physical structure here too. Now we begin to see the clear differences in front and back and in left and right. However, the blueprint of this layer is pretty simple as there is only one line of development in the coding, which means that differences in physical shapes will not be as varied as in the previous layer. Again, mammals will develop in water as well as on land. However, the structure of the mammals becomes more and more complex, which creates difficulty in structuring varied mammal life in water. One of the specific developments is a more efficient mechanism for the uptake of "light" energy, the breathing system, so that the animal has access to combustible fuel from gases. As water contains less air, mammals who live in water cannot completely satisfy their need for oxygen and they have to, now and again, "come up for air". This limits the development possibilities in water.

The fifth day can be described by 5.5, 5.7, 5.2, 5.6, 5.4, 5.1 and 5.3.

These are the seven phases of day 5.

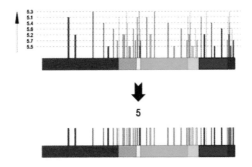

5

The 6th Day: Human Beings

The code for this energy layer 7 is 7–5–2–6–4–1–3. This represents the creation of the human being.

You may notice that the blueprint for the human being does not differ much from the one for mammals. They are basically structured alike and they function alike. Development in water is no longer an option because air energy has become the main energy supply source, as opposed to digestion. This means – and I am really sorry about this – that mermaids are just a fantasy, although they are a possible fantasy! Energetically, humans exist in water but the expression in a physical format is not possible. The physical likeness between the highest mammals, say the chimpanzee, and the human is an expression of the code likeness. However, this is going to change in the future. The chimpanzee has finished its shape and development while the human being is only at its cradle, which is still very close to the top of the previous layer. Development for the human being lies mainly in the nervous system, which will have an impact on the shape of the body, and also in the head structure.

The sixth day can be described as 7.7, 7.5, 7.2, 7.6, 7.4, 7.1 and 7.3. These are the seven phases of day 7.

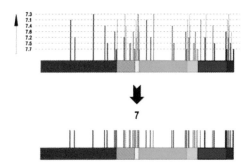

The 7th Day: ?

The code for this energy layer 3 is 3–1–4–6–2–5–7. This represents the creation of … At the moment this is still a question mark because day 6 hasn't been completed yet and therefore day 7 cannot manifest yet.

The seventh day can be described as 3.3, 3.1, 3.4, 3.6, 3.2, 3.5 and 3.7.

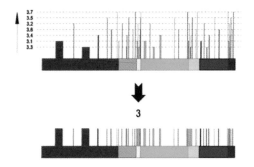

The following drawing shows how the spectrum of visible light has been coloured in day by day, starting from the seed 3, creating 1–4–6–2–5–7–3. Here we have shown the code 1–4–6B–2D–5–7–3.

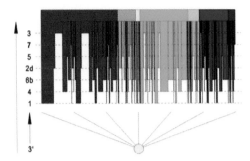

The various possible steps of the eight possible codes are represented in the following structure. The thicker line shows the development towards the human being.

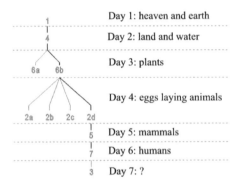

Simply by following the energetic lines and by recognising what those energies express in physical forms, we can gain an overview of all living matter on this planet and in our universe. Everything we have discovered so far needs to fit into this framework and there needs to be room to place everything that we haven't discovered yet into it too. When we categorise living things in this way then we are able to view similarities and differences straightaway. This simplifies our understanding of physical structures, what is possible within the physical realm and what isn't. Once we have this framework, people can occupy themselves by filling in their own field of expertise into the energetic codes. In this way our world and all of creation will become a recognisable place. We will be able to understand the behaviour of plants and animals straightaway and where they belong in the greater scheme of things.

On a smaller scale, every individual can enjoy searching their own lines of energy within their own little world, leading to an understanding of the development of their own self, their family and even society as a whole. There wouldn't be a need to judge events any more from a single narrow viewpoint as we would be able to understand the larger picture and the relevance of this in evolutionary terms. This would lead people out of the never-ending beliefs of "good and bad" and "right and wrong". There would be no more fighting between head and heart, between logic and intuition. This is the freedom that Asians call "Nirvana" and Buddha called "Enlightenment".

The Story of Creation

Our study has resulted in a code for the universe and its develop-ment, which now allows us to tell the story of creation against this backdrop. But we need to start at the beginning, when there already was something! Nothing can be created out of nothing. There has to be a base layer in existence in which everything else can manifest itself. One cannot even create visible white light unless you have a medium for the wave to travel through, a medium in which the waves can interact. When there isn't a medium, the light remains unnoticed, invisible, not manifest. So you need a seed to develop something, but you also need soil – a fertile environment.

The potential to manifest everything is present in the environ-ment, in an already existing layer. We know the existence of such a layer. Space, as scientists tell us, is vibrating with potential where what they call *Virtual Particles* are shooting in and out of existence in an almost undetectable way. This part of our universe is called the **ether**. Einstein referred to it as the ether and in Ayurveda it is also known as the element ether.

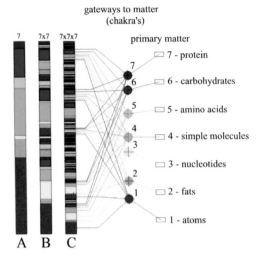

The ether exists just like all other physical manifestations out of the seven main layers, ranging from very light to very dense and compact. The known configuration can be written as 1–4–6–2–5–7–3. When the pressure on frequency 3′ becomes so intense that the energy field can no longer contain it in its current form, the frequency "collapses" inwardly and manifests its seven layers, in the same way we have demonstrated with visible light. This implosion is recognised by our scientists and is called the Big Bang, which in Ayurveda is referred to as the element **fire**. Now this fire, the result of the compacted energy, escapes and spreads through the ether and meeting the ether, which will alter the characteristics of pressure and temperature in certain places of the ether. The seven frequencies that emerge from frequency 3′, from light to compact, we already know. They are 1–4–6–2–5–7–3. Because this interference results in certain places in the ether becoming denser, these places become more manifest, which kick-starts the creation of the universe in its physical form. It forms a nebula, a gas cloud, in which, step by step, the seven layers of *Primary Matter* will manifest; seven layers of the smallest possible matter. The gas cloud is described as the first physical expression after the Big Bang, and is known in ancient texts as the element **air.** This primary matter also splits into seven other layers, the heaviest of which takes a while to create matter. These layers within the gas cloud, within air, create matter as a development of the first layer of the creation of the universe. In other words, it is a detailed development of that first layer, which in the code is represented by frequency 1. The internal structure of 1 we already know from the division of the colours as 1–3–4–6–2–5–7. Primary matter is, in the first instance, an array of atoms. We have a table that organises all the atoms in the universe, the table of Mendeljev: 1 halogens and inert gases, 3 non-metals, 4 metalloids, 6 post-transition metals, 2 transition metals, 5 alkali earth metals and alkali metals, and 7 lanthanides and actinides. Once this layer is completed, a new one will emerge within the universe.

When the 7 frequencies of fire (7 components of frequency 3) move through the denser areas of air, frequency 3 presses the atoms to become more compact, to take up less space. The pressure in the gas cloud increases and, so to speak, pushes the matter even closer together. There is simply less space available for matter to occupy, which results in an even more compact and better structured matter. This phase is known as the element **water**, which contains a more organised material manifestation. The atoms join up to become molecules. This second layer of the creation, frequency 4, shows us 7 densities in molecules too, which are the simple molecules (NaCl, H_2O). Many of these are light enough to remain in gas form. These can combine in a great number of ways to create larger structures, which we come across in our material world, such as water itself or salt crystals within the smaller range, and other heavier molecules that are soluble in water.

And, again, the pressure increases on these material structures, resulting in a further "hardening" of the matter, which eventually leads to such a density that the moving creative wave of the universe can no longer penetrate. This impenetrable layer of the creation has been named the element **earth**. Here the matter will be compressed to such an extent that no further space-saving structural reformation can be done. In frequency 6, all heavier and more complex molecular structures are formed, some of which are still soluble in water but which can always be isolated as a solid. These solids evolve throughout the layers of frequencies and they can be grouped as: 6 carbohydrates, 2 fats, 5 amino acids, 7 proteins, and 3 nucleotides.

Then, under the ever-increasing pressure (and the cooling down) a further development takes place. The gas form, frequency 1, remains, whilst part of this creates frequency 4 manifestations, the liquid form, which also remains. But the solid form can absorb the pressure only by rearranging the matter, because the frequencies cannot move through this layer and the result can only be more compression. This creates different structures within the solids. At this level, frequency 6 development, the solid pieces will join up to manifest the structures of tissues, arranged in the most economical way, so that the matter occupies the least possible space. Together they create a cell!

In this way the matter is now capable of recreating itself. Energy still flows in, but because the cellular structures are too compact, thereby obstructing the flow, the structures begin to swell up. In frequency 6, the energy flow creates an organism. The structures *Explode* outside the framework, in a way like the original Big Bang. The substructures need more space to absorb the pressure and they enlarge, they expand, like a balloon that is being blown up. This organism has two sides that look pretty much alike, the roots and the plant itself, an upper and a lower side; two parts of one whole. The tissues of the plant are, in principle, the same tissues as those of the entire universe because they originate from the same energy layers: 1 juice, 4 resin, 6 softwood, 2 gum, 5 bark, 7 leaves, and 3 flowers and fruit.

The first cells are formed within the upper layers of frequency 6. These are the primitive life forms that scientists are looking for in different places of the universe. Single-celled organisms, from which others emerge, are bacteria, amoeba, plankton and so on. Frequency 1, the first layer of the developing universe, is responsible for the atmosphere – the gas formations. Because of increasing pressure and/or decreasing temperature, the energy inside every layer solidifies more and more. This manifests itself in the next seven layers of density. When frequency 1 can no longer be compressed further, a new layer emerges from the bottom of this frequency: this is frequency 4. This layer is responsible for the fluids and it develops seven different manifestations, until out of the last layer a new frequency shows up; frequency 6. And, again, seven layers evolve within this frequency until the pressure can no longer be contained unless a cellular structure is formed.

From here, the increasing pressure from the inflowing energy can no longer be absorbed into an increasing density of the structure of matter. From here on, the cells are forced to replicate and the matter grows in size and complexity as the seven layers develop more and more, because within the physical structure of the organism every new energy layer that absorbs pressure develops the organism frequency by frequency, layer by layer, type of tissue by type of tissue. From frequency 6 onwards, these cells can grow because the inflowing energy of the seven layers of frequency 3 gets held in primary matter, which will expand into primary systems. These are, in the first instance, organised as plants when they enlarge in water as well as on land. They condense their tissues in a seed form, which is the most condensed manifestation of the element earth. A plant will balloon out from its specific primary tissue information, contained within the seed.

Once frequency 6 has fully manifest itself, and the creation of the seven different layers of plant life has finished, the still inflowing frequency 3 will awaken the next layer of the earth element, and that is frequency 2. Following this development, we can clearly see how the primary matter of a cell can be enlarged in every frequency until seven primary systems become apparent. In the scheme below, we complete the lines between the frequencies and the primary systems:

- 1 form   water   lymphatic system
- 4 balance   blood   circulation system
- 6 consciousness   muscle   respiratory/digestive system
- 2 movement   fat   motor system
- 5 communication   bone   sensory system
- 7 knowledge   nerve   nervous system
- 3 personal power   seed   excretion/glandular system

These manifested systems organise themselves in all organisms in a way determined by the energetic code. In frequency 2, there are 4 different combination possibilities within the energy field, which result in an enormous number of animals that have one thing in common; they lay eggs to procreate.

First, we encounter the invertebrates with a primitive nervous system. This is the first layer of the 4 possibilities within frequency 2. The code for these, as we already mentioned, is 2–6–5–4–7–1–3. Invertebrates with the beginning of what can be deemed a central nervous system, a separate brain, are expressions of the code 2–6–5–7–4–1–3. Gradually, the amount of bone tissue increases and the vertebrates become fact. There does exist a layer of vertebrates with a primitive nervous system, however, coded as 2–5–6–4–7–1–3. From these, only a few species are known in the entire world. The great majority of animals that manifest, and that have managed to establish themselves a place on an ever-busier surface of the earth, are invertebrates with a primitive nervous system, some with a very simple primitive central nervous system. The vertebrates with a complex nervous system, with a clearly separate brain, are coded as 2–5–6–7–4–1–3.

Once the egg-laying animals have developed and they have found a balance, the next density layer receives the energy and begins to grow. The dinosaurs are completely extinct because their design was too large, their vegetarian food requirement gigantic, and their natural predators too few, which resulted in the survival of only a few smaller examples, of which the crocodile is one. After this density layer, mammals arise from frequency 5. And once this layer has been completed, the time is ready for mankind to enter the stage out of frequency 7.

The changeover from plants to egg-laying animals, to mammals and to humans are all steps in the development of the universe where Darwin's rules of evolution, about the specific influence of the environment on the selection of the species, is not applicable. Humans are not direct descendants from apes in the same way that a rat is not a direct improvement on a bird or a fish. These can all be described as separate steps in which a "new" layer of potentiality is opened up, a layer that then physically manifests and the new developments actually happen in a field that sits perpendicular on top of the already existing, previous, one. Darwin's rules described the forces that are of great importance in the development of species within each layer, but they do not explain the structural changes that happen between the various layers. The explanation is that every layer expresses a different energetic code.

All animals have the same primary systems, but they are manifested in very different combinations, which results in very differently structured organisms. Mammals have wombs and facilities to feed their offspring; structures that more primitive animals do not have at all. Human beings appear to be so close, in structure, to the most highly developed mammals only because human development is still in its infancy and the energetic codes are so similar: 5–7–2–6–4–1–3 for mammals and 7–5–2–6–4–1–3 for a human being. We are just emerging and we are infantile and naïve in our learning process about our place in evolution. Often our behaviour is very animalistic, but through evolution we will become something totally different. Human beings are manifestations of frequency 7 and the content of this energy is knowledge, which will show itself in the development of the nervous system, especially the central nervous system. This also means that our physical appearance in the future will change too, but until then we are lookalikes of monkeys.

Let us then become conscious of the growth process that we are part of, and conduct ourselves as real students with respect for teachers and the school. We are attending the School of Life and we should be respectful of everything that surrounds us and that is part of this evolutionary process. This includes the world and all that is manifest within it, even the people we don't like very much. Just remember that those are the people who teach us the most about what we are supposed to learn, and that those who show us that destruction, greed and power are not courses of this curriculum are important too.

Chapter 5
Frequencies and Matter

Imagine the whole universe, and everything that has been created within it, being held in a seed, encapsulated and perfectly still. There is no movement and all is in perfect equilibrium. This is your Garden of Eden, where everything is perfect and nothing ever changes. There is a perfect stillness across the whole fabric of the universe. There is nothing in it, except enormous potential. Everything could sprout from it, but as long as nothing changes nothing actually will sprout. Consider it to be the soil in your garden but without anything emerging from it; the seed in the ground remains dormant. It is barren land, but it is also full of potential; waiting for something to change. The best image we have of this is volcanic lava; the most fertile ground on earth, but which lays barren for a very long time.

And then something changes! The seed cracks open and a tremendous burst of stored energy starts to spread out from that point. The pent-up seed's energy explodes into the area surrounding it, thereby disturbing the fabric of that area. The Big Bang has occurred.

Inside the seed lies information that is stored in 12 different layers: 7 frequencies (notes) are responsible for the creation of the "observed", and 5 frequencies (semi-notes) contain supporting information. To move one note through an octave and to end up with the same note but with half the frequency, or, in other words, the same information but in a different package, there are 12 steps. There are 13 notes (full notes and semi-notes) within the span of an octave. A scale is comprised of 8 notes, of which the fifth and third note create the basic foundation of all chords. Note how the piano keyboard scale of C (third note) to the C above, of 13 keys, has 8 white keys and 5 black keys, split into 1 group of 2 and 1 group of 3. Ratios found in the first 7 numbers of the Fibonacci series (0, 1, 1, 2, 3, 5, 8) are related to key frequencies of musical notes. As a wave commences its journey from the burst seed, step by step the information from the seed takes form in the wave.

The first frequencies of the initial wave that leaves the seed (after the Big Bang) are the frequencies that make up the octave of light. *First, There Was Light.* Looking at the position of the light spectrum within the electromagnetic spectrum, it is a very small band somewhere in the middle of the whole spectrum. The indication is that we haven't quite been able to measure the furthest reaches of the universe. Light is, in other words, the seed spectrum from which the whole universe is created.

As the wave starts to make its way into the "space" around the burst seed, lots of energy is radiated outwards. The undisturbed coldness from space – when there is no movement, there is no heat – is set into violent motion by the energy wave that passes through it. The energy of radiation is movement energy (kinetic energy), some of which can be transferred to the formation and movement of a particle. Its motion will be in a direction perpendicular to the travelling wave; the particle jumps up and down, as the wave spreads longitudinally and donates part of its energy to this movement. The further the wave travels away from the source, the more energy is transferred into the motion of the space fabric, and the less amplitude is created in the movement of the particles as a direct result. So, the further away one gets from the source, the less movement is transformed into particles. The frequencies remain the same, so the same matter is created, only in a lesser quantity.

Also, the longer the source sends out its information, the more heat is released and the slower it will release the information, because the temperature drops as the energy within the seed diminishes. Here, the frequency of the information wave itself gets less as time moves on (wavelengths lengthen). So, the combination of denser matter through which the wave travels (creating matter on its way), and the slower information wave emanating from the seed, results in the manifestation of all possible seven layers within every seven layers of space (octave after octave).

A wave that passes through denser matter will remain at the same frequency, but the wave speed will slow down. This allows for the frequencies that make up the wave to move through the denser material at different speeds, and this shows itself in the separation of the information into its seven components. A beam of light passing through outer space is not visible; it is very dark in outer space. The reason is simply because outer space is loosely populated with particles; it has a very low density. If light is allowed to pass through a denser matter, such as air, then the light beam shows up as a solid white beam. Pass it through a water droplet, even denser than air, and the colours of the rainbow appear as individual bands from the solid beam that entered the water. In these three circumstances, the frequency of light has not changed, but the speed of travel has, and the denser material "filters out" the frequencies otherwise hidden within the overall frequency of the whole.

Light becomes manifest when the information travels through a certain density, and it shows the information in greater detail when the medium it passes through is even denser still, until it eventually can no longer pass through the densest environment. When it shows its information in the broken-down form, this is the creative process within that particular layer of density, in this case the colours. The colours were always there, but now they are manifest. They have been created in the outside world, in space, while the potential for that creation was already held within the information.

As the wave passes through space, disturbing the quiet peaceful fabric of space itself, it starts to manifest the information it carries. There are seven main "notes" within an octave of frequencies; these are frequencies that will directly manifest and form matter in a certain way. So, there will be seven different layers of matter. The first disturbance of space has created wide bands of potentialities as we move away from the blast site, with each new band being denser than the previous one from which it was created.

The seed is, as mentioned, light. This light separates "heaven" from "earth". "Heaven" means the undisturbed space where nothing ever changes and everything is perfect, and "earth" means the object in motion, the unfinished creation. That is your day 1 of Genesis, and it is the first band in which the information contained within the electromagnetic spectrum will be manifested. Here we will find all possible frequencies, being a wave, but also a particle, depending on how we observe it. Here, too, we find all potentiality, and from here all transformation begins.

On day 2 we separate the water from the land. In other words, instead of just having gaseous clouds where there was previously nothing, only sheer potential, there is now water (fluid) as a manifestation of a cluster of atoms (particles). This layer materialises the information into larger molecules and conglomerates of interactive particles.

On day 3 these particles have organised themselves into more complex structures, which we call plants. These structures start to have a solid base and they tend to stay more or less put, certainly the later developments. Plants occur on both parts of the already created substrata, water and land. Plants are the outward radiation from the soil from which they spring.

On day 4 we see the creation of the first animals, which are characterised by the fact that their procreation involves the laying of eggs. So we are talking about the fish, the reptiles and the birds. The creation of these bodies, the formation of the skeleton, happens in a plane that sits perpendicular upon the already existing manifestation.

Day 5 sees the introduction of the mammals, again both on land and in the water. At this level of the creation, the senses develop on top of the structure they received from the most highly developed animals from the previous layer. Again the new development, the senses, manifest as a detailed structure upon the existing one: big ears, big nose, an enlargement of the face.

And on day 6 human beings are created. Here we expect to see, eventually, an enlargement of the nervous system. This manifestation will also sit perpendicular upon the existing skeletal structure, which allows us to imagine a front-to-back head development.

Six different creative bands progressively appeared moving away from the source. Each band has its limitations of expression due to the density of the matter in which the information is being expressed. Each band is created out of the previous one, and therefore contains all the expressive potential of the previous band plus its own. The information within the original wave frequencies, seven major areas (plus five that won't manifest into matter), is in the second band, the creation of land and water, only expressed in the different layers of matter.

Human evolution has only just begun and there is still a lot of information to be expressed before this band will have fulfilled its potential. Only then will the seventh and last band be formed. Hence, we have no idea what that might involve or look like. So let's not speculate.

Each of the seven major frequencies holds specific information, and the way this information appears in its interaction with the space fabric the wave travels through shows the progressive densification of the materialisation of this information as the layers manifest. It all bursts open from a seed that holds power, in which the pressure has become unbearable.

In the manifestation of matter, the first frequency that appears, note 1, fixes the energy in matter; we are **Grounded**. Now we *have* the experience; now it has become "real".

The second frequency to manifest, note 4, delivers **Balance**, a balance between attraction and repulsion, the love–hate balance; me and the outside world.

The third frequency to appear, note 6, holds information about **Imagination and Consciousness**. It compresses knowledge into what we believe it to be. Jesus, Buddha, Mohammed, they all possessed knowledge that they shared with others. Other people "believed" this to be certain things, which led to a division amongst "believers". Each of these groups believe they are the rightful owners of the knowledge these people shared.

The fourth frequency that appears, note 2, tells us about **Desire**, and brings desire in contact with knowledge, which gives us a first glimpse of how things could be.

The fifth frequency that appears, note 5, handles **Communication**. It explores the inner self, the personal authority and the way the self is expressed. The knowledge will now be passed on to the deeper layers of the creation.

The sixth frequency that manifests, note 7, contains information about **Insight** and **Knowledge**. Without this no goal-oriented creation can become reality. Before creation begins, this knowledge has to be present and will arrive on the surface right at the end of creation.

And then the last frequency is the same information frequency, note 3, as was present within the seed, only the manifestation then was one of an octave higher. It now contains all experiences and gathered knowledge from the previous six developments. We are now ready to sprout another series of seven, in which essentially the same information will be manifested but in a slightly different, and more developed, form.

Ayurveda teaches us the formation of elementary tissues out of these seven bands of information. According to this ancient knowledge, all tissues in this creation come from the densification of the previous layer; one comes forth from the other. The plant, like the human being, and the universe itself, is similarly composed of seven planes.

The variety of plant life we see occurs because the third day layer will divide up into another seven layers of sub-densities, which will diverge the information into another seven different details, all recognisable as plants but grossly different in plant appearance. Furthermore, these seven layers of fabric density will separate out into another seven different layers each, extending the possibility of diversification almost to infinity. For instance, the plants grow in seven different levels: tall trees, low-level trees, shrubs and bushes, vegetables and herbs, mosses, root vegetables, and last but not least climbers.

A human being is made manifest in a similar way but not on day 3. On day 6 the human being appears on the scene in the story of creation. The fabric of space has cooled off even further; in fact, it "cooled" three more stages. The first creative wave from the source is now displaying the information of the first frequency (layer 6 of creation), which is about grounding. It is also about the highest density, apart from the seed itself, that the information within one octave, one layer of creation, can get. It represents the nervous system of creation; in other words, the way to become aware of experiences and note them down. That is the function of the human being within the evolution of creation: *To Make Creation Become Aware of Itself.*

On day 6 a new expression of these seven frequencies becomes manifest. This creature also has seven tissues, just like the earth itself, the plant and the animal before it. Here we recognise plasma (note 1), blood (note 4), muscle (note 6), fat (note 2), bone (note 5), nervous tissue and bone marrow (note 7) and reproductive tissue (note 3). Of course, the first manifestation of these tissues is at a cellular level, and it helps that New Biology has now proven that every cell in our body has all the same organs as the whole body itself. These are made up of rudimentary tissues that are the same as the plasma, the blood, the muscles, the fat, the bones, the nervous tissue and the reproductive tissue we find in the complex beings that we are. The cell is the micro-cosmos of the entire body, a holographic image.

Ayurveda teaches us that human beings are made up of three essential sets of qualities; each specimen in a different combination. The combined attributes of *Dry, Cold and Light,* together with *Irregular, Mobile, Rarefied and Rough* make up the quality of **Vata.** A combination of *Oily, Hot and Light,* together with *Intense and Fluid* gives us the quality of **Pitta.** The third is a combination of *Oily, Cold and Heavy* together with *Stable, Dense and Smooth,* and is called the **Kapha** quality. These three sets of qualities combine to give us seven different human types. They represent the seven "major notes" on the human scale, whereby we can make the link to the seven major frequencies and the information they carry. Each human type then corresponds, in its qualities, to the manifestation of mainly one of these major frequencies, also creating denser, more solid, human beings as evolution progresses. Each stage of the development of human consciousness begins by frequency seven (insight) and ends with frequency one (grounding) to sprout again from frequency three (power). Each time there is a major shift in human consciousness, there is the turmoil of changing into another octave, getting into another gear. It is the moving into another density of the space fabric. As an example on a larger scale, the post-war years (1940–1945) showed an explosion of materials and a materialistic way of living, to the extent that our churches became emptier. Right now, as a result of that development, there is a massive interest in spirituality and an explosion of awareness in which we find the condensed knowledge that all human beings have experienced so far.

If we were to describe each of the seven human types in one word, which is, of course, a serious simplification, we could see how the types relate to the information from the various frequencies. In the manifestation code, 1–4–6–2–5–7–3, we notice that the tissues become heavier as we move from 1 through to 3.

- The first frequency has information about **Grounding** and about **Surviving**, which we find in the Kapha–Pitta type.
- The fourth frequency has information about **Balance** and **Fire**, which we find in the Pitta type.
- The sixth frequency has information about **Inspiration** and **Consciousness**, which we find in the Pitta–Vata type.
- The second frequency has information about **Movement**, and that is to be found in the Vata type.
- The fifth frequency has information about **Communication** and **Togetherness**, which we find in the Vata–Kapha type.
- The seventh frequency has information about **Knowledge**, anchored in the Kapha type.
- The third frequency has information about **Personal Power** and it holds all the potential within, just like the Vata–Pitta–Kapha type.

Energy is everywhere and all energy fields seemingly exist together in a melting pot of potential, although it is difficult to imagine how so many different things can all be in one place at the same time. Even when we accept that energy doesn't necessarily take up physical space, we still have the problem of visualising how so many different waves can be present in one spot, because we "see" the movement of the universal particle, up and down, which can only be one particular wave at the time. They may change quickly, but at any given moment in time we only observe one particular motion; therefore, only one particular wave expression. Yet we know that the number of waves in any one spot at any given moment is infinite.

The "infinite" bit refers to the long sequence of subsequent layers of seven manifestations that we recognise. Twelve steps conclude an octave and at the same time start a new octave in which we encounter the same information, organised in the same way, but with half the frequency. As the wave emanates from its origin, the wavelengths manifest in very specific quantities as time goes by. It "loses" energy; it "donates" energy to the environment, the fabric that it travels through. It also vibrates less energetically, whilst holding the same information. By creating an environment that is slower, denser, all subsequent information input will manifest in a denser way, step by step separating out what wasn't noticeable (visible) within the high-frequency movement of the original. This is why light (an electromagnetic wave) emanating from any source in a virtual vacuum like space itself, is not even visible. It vibrates at such a high speed, hardly interacting with the rarefied medium it travels through, hardly disturbing the fabric of space. Where space has already contracted, i.e. become denser, it separates the visible light as a well-defined beam from radio waves, x-rays and the other parts of the electromagnetic spectrum as observed in our atmosphere. With each level of density that light encounters, it shows more specific characteristics of what was hidden inside the wave, the hidden harmonics within the overlying frequency. By slowing the travel speed of the wave down, it smears out finer aspects and details of the information held within the wave. As the wave, which originates at the source, then travels through the various media, it delivers more and more details of manifestations into each medium.

When we examine the information held within the main seven "notes" of the universe, and how this information manifests at different levels, we can begin to see not only what is being created, but also why certain things manifest in that particular way.

From what we have said so far, it should be clear that even when we talk about seven layers of "information", these also need to manifest in a very specific order, from least dense to most dense. From this most-dense layer then springs a new "octave", which contains the same information, but manifests it at a different frequency.

So, all manifestation comes from frequency information that we can describe as *emotion*. From this seed we need to identify seven main layers of information, whereby each layer becomes "more dense" and the whole structure of these layers forms a complete entity, such as the earth, an electron, a protein, a plant, or a human being. The seven layers, with each layer containing specific information and each an expression of this information, forms the structure, the physical manifestation of the information. We can identify the seven layers of frequencies, increasing in density as we move forwards, as *form, grounding (1) – balance, love (4) – imagination, consciousness (6) – movement, desire (2) – communication, togetherness (5) – divine knowledge (7) – emotion, personal power (3)*.

We have already spoken about the two contrasting forces within our universe, that of the strong force and the weak force, which are opposites; one attracts and one radiates out. We have talked about the electromagnetic force and gravity; one radiates out and one contracts. It is this interplay between such forces that is responsible for vibration in general, creating movement past the point of equilibrium. The principle hidden within those forces is the principle of yin and yang, contraction and expansion. Every frequency has a higher aspect – a looser part, the upper part of the sinusoidal curve, moving it upwards – and it has a lower aspect – a denser part, the lower part of the curve, moving it downwards. At the same time for all matter (yin) there is its anti-matter counterpart (yang), which is now being acknowledged by modern physics.

Every frequency also has those two parts – upper and lower – and these can be identified within each of the frequencies:

Form has two aspects: the stability of the structure (yin) and the flexibility (yang).

Balance has two aspects: attraction (yin) and repulsion (yang).

Consciousness: intuition (moves from inside out – yang) and insight (moves from out to in – yin).

Movement: contraction and pulling towards (yin) and relaxing and pushing away (yang).

Communication: talking (yang) and listening (yin).

Divine knowledge: in our actions (yang) and in our thinking (yin).

Personal power: held quietly inside as certainty (yin) and passionate behaviour (yang).

Within each frequency, these two principles are constantly at work, ensuring that everything within our universe is in constant vibration and movement. Since the Garden of Eden, Nirvana, or the seed from which this creation has sprung, nothing has remained unchanged. It will take a long time for this universe to stop vibrating, for the ripples to die away and for everything to come into a perfectly still balance once again.

Chapter 6
Creation of a Human Being

When Mummy and Daddy Really Love Each Other and Come Really Close Together, One Little Sperm from Daddy Enters a Ripe Egg from Mummy and a New Life Starts.

They tell us this new life will have a lot of the characteristics from both the mother and the father because it receives the genetic information from both. The sperm brings a copy of the father's DNA and puts it in the same bowl as the copy of the mother's DNA that is already contained within the egg. And here already is the first of many problems with the conception story that we grew up with!

Every living cell has, as we know, a double-stranded DNA in its nucleus. The reason for this, we are told, is that the cell divides into two, which it does in order to propel life into the next generation of cells. In doing so, it helps to maintain the life of the tissue itself and the two strands unravel. Each of the two new cells will get one single strand of the original DNA. These two strands are exact mirror images of each other, which means that the two new cells, each receiving a single strand, can build the complementary strand to the one they received, in order to complete their full DNA information. Each of these two new cells now contains exactly the same genetic information as the "mother cell".

We are told that the first human cell is created at conception when the sperm delivers the single DNA strand from the father into the egg cell, where it combines with the single DNA strand from the mother. This process completes the coming together of the genetic information of the parents. Except, of course, for the fact that these two single DNA strands are not identical! They come from different people with different genes, and therefore these two strands cannot form a complete double-stranded set. They just don't fit together.

Although it isn't specifically said, it is implied that it will be all right because for every gene the cell now has the choice of which one it will copy. The new child might have mummy's eyes, nose and legs, and daddy's ears, chin and belly button. This is not acceptable on a variety of levels, but, scientifically, foremost is the fact that we have known almost right from the start of the Human Genome Project that the theory stating that there is one gene for every protein the body needs, is completely wrong. There is no single gene for any physical expression at all. A single genetic code, for instance, does not determine the colour of the eyes. This then means that the first cell cannot have made a decision as to which code it should copy in order for the baby to definitely have mummy's eyes.

In the "official" story there is also no mention of who is actually deciding which gene to copy. Who is in charge of this decision-making process? The mother, because she is hosting this event? The father, by some remote-control mechanism? The child who doesn't exist yet? And please don't say it's pure chance! Or the luck of the draw. The more we find structure in the chaos we find ourselves surrounded by, the more it shows us that chaos is something we fail to see the structure in. Similarly, chance is something we fail to see the logic in. When we don't recognise the links and the particular influences that make it what it is, we call it "pure chance". This says more about our own ignorance than about how things truly are.

What really comes together at conception is the male (yang – weak force) and female (yin – strong force) energy with the power (fire – electromagnetic force) to create a new life. It is the willingness and openness that we call love that is responsible for the potential this situation holds. You may argue that there are circumstances in which a pregnancy occurs when at the time of conception this willingness and openness is not present. You may argue that, but the truth is that we don't know. No matter how bizarre the world sometimes may look to an ignorant mind, it would be a mistake to judge nature before we understand it. Let's simply pursue the common line in order to learn a bit more about the reality of the creation of a new human life.

At the moment of conception there are male and female issues that will dominate the ether surrounding the two people involved. The cell, which is the physical manifestation of what we are talking about here, has now been completed with all of the information it needs for it to start working on the jigsaw puzzle. In the first moment there is no definite human presence; there is a lot of potential though, within the human field. Gradually, the choices are made and a specific picture, in genetic terms, will start to emerge. Who is making the decisions? As always, in every situation, it is the *environment* that determines which way the potential development will go. A corn seed has the potential to develop into corn, but whether or not it will depends on the soil the seed has fallen on. It may lie on a rock and never develop at all, or it may sprout in very poor soil and become a weak plant, carrying little fruit, or it may fall on rich soil and become a healthy and strong specimen. Furthermore, it is dependent upon rainfall, sunshine, storm winds and many other environmental factors that will shape the growth and development of the seed. It is no different for the development of the human cell. It contains all the information but whether and how this will be deployed depends on the environmental factors. And these are, as we now know, also the energies surrounding the primal soup of this human cell.

With all of the ingredients in place, the first task is to form the double-stranded DNA of this new cell. The energies flowing from the parents towards each other, and therefore inside the mother, will be translated into a physical code that will form the basis on which the later structural development will take place. Once the first cell has been prepared with the cytoplasm and its various structures, as well as with the nucleus and its genetic material, we have the blueprint for the organism as a whole. New Biology has taught us that every cell has all the same structures as the completely developed organism, which means that there is a digestive system, a breathing system, circulation, a musculoskeletal system, a nervous system, an immune system and so on. This new unique cell contains the information on which the development of the whole organism will take place. We are now ready to divide and start the building process.

In embryology books it tells us how that first cell divides into two cells, then four, eight, sixteen, and so on. Soon we have a small cluster of cells, which forms the very first stage of foetal development. The next picture in that story is one of a small water-filled balloon with a thin layer of cells stretched out on its surface. Nobody explains how we get from a cluster to a layered cellular structure, but that is how it develops. We could say that the first stage indicates a separation between the mother tissue and a new independent tissue, a cluster of cells that no longer belong to the mother tissue. In the second phase, we see the separation of the solid matter from the fluid; separating the waters from the land. In the Biblical Story of Creation, we have reached day 2.

At five weeks' gestation, the foetus takes on the shape of a bean, thereby representing the plant-life stage of day 3, and showing the morphological signs of a primitive animal by week six. Day 4 gets us to the seventh week of gestation and the foetus shows as a fish shape, while by week eight we can definitely recognise a monkey (day 5) in our human foetus. Day 6, in the ninth week of foetal development, we see a human form appearing, which becomes more obvious between the 10th and 12th week of gestation. From then on, this form will continue to become a more recognisable representation of the human form, as we know it, until it appears complete at around 36 weeks. This completion also includes the fact that the females already have all the eggs inside their ovaries for the rest of their lives. Therefore, it is important to realise that the human seed is also totally completed by the time the human baby is born. In the male form, a similar development has taken place; however, not all sperm is in situ in a physical form. Nevertheless, energetic development has concluded and no changes will be made to the genetic content after this completion.

Viewing foetal development in this way, it becomes obvious that, during the first 12 weeks, we run through the entire universal development in a rudimentary fashion. Every single major step this creation has seen is recorded within our DNA. In order to begin developing as a human, we first need to get to the point where everything is ready and in place so that human life can become a reality. Some religions believe that the human soul descends into the foetus around week 12 of development. Scientific statistics show that most spontaneous abortions occur between week 10 and 12, and it is thought that this is due to there being something wrong with the foetus to the extent that life could not be possible. Indeed, a human life would not be possible if we could not develop beyond the 12th week. The Human Genome Project has thrown up another interesting fact in that geneticists have stated that 93% of human genetic material can be classified as "junk DNA". What they mean by this is that we apparently only need about 7% of our genes to be able to function as a human being. This begs the question: "Why don't we shed this useless baggage?" The answer may well be that we do not use it *To Function* as a human being; we used it *To Become* a human being.

One aspect that is probably not supported by this "foetal development running through the universal development" theory is the fear that medical authorities have spread about when a woman

drinks and smokes during pregnancy and the likelihood that
the baby will suffer abnormalities as a result. From observations
we know there is truth in that concern, but it doesn't actually go
together well with the medical "knowledge" that says that the
first three months of foetal development are the most important
because the foetus then is most vulnerable to disturbing influences.
Often we hear women express these fears when they have had a
boozy weekend whilst not realising that they were pregnant. In
those early pregnancy stages, the damage to the foetus then should
be irreparable, and yet it never even shows up. The reason for this is
that the development of the foetus, at that stage, is virtually totally
independent from the outside world. You would need to almost
destroy the mother – the outer environment of the foetus – before
the foetus would be affected. The first three months is the most
protected stage of foetal development, not the least protected! Of
crucial importance for that new life is the energetic feeding ground,
which is how the energies of the mother are flowing at the time
of conception.

The first three months of foetal development shows the history
of the universe's development in the same way that pack ice at the
North Pole holds a record of the weather conditions for many thousands of years previously. The compact ice is a specific compressed
form of the main elements that were present at the various stages
throughout time. Expressed in the larger picture, true conditions
at any given time can be recreated from the compacted information laid down in the material, in this case ice. The history of the
universe, right back to its initiation, is written down in a material
code, which we call DNA. That is also why every living cell, even
a bacterium, has genetic information, because without that information it is not possible to create something as complex as a cell.
And as cells are used to create larger and more complex life forms,
so they will need to follow written guidelines in order to get to the
point of evolution that that strain has already reached. This point

will be the starting point to gather more information, so evolution can move on. In other words, creation has not finished yet!

Once the human stage of development has been reached, the cells that are fully developed as human cells can start their development into a human being. This means that from this moment on they will be fed by "human information". Now, it is important to remember that all life is energy, and that all food is energy. If we add that to the equation it means that the cells now pick up energy that has specific meaning for humans. But what does this mean in real terms?

New Biology has confirmed that the surface of all living cells is covered with little antennae through which the cells communicate with their environment. It is via these antennae that messages from the outside get relayed to the inner world of the cell, and vice-versa. Only it turns out that the antennae don't operate through physical contact of proteins or hormones. Almost the entire function of these antennae is energetic. They each pick up a specific frequency, which is then transferred and translated into an intracellular process. Our cells listen to radio signals in their environment in order to relate their activity to the environment they live in. In other words, just as our own grown-up system gets its information from the vibes of the surrounding area, so do the foetal cells get their information from their living environment. In a way, we could say that there are no surprises there then!

The foetal cells receive information from their environment, which is the mother carrying these foetal cells inside her. Hence, the maturing of these cells is entirely dependent upon the information they pick up from the mother's internal system. By programming their own system by the way the mother operates, they prepare themselves to enter a world they know nothing about. The only information these cells, and therefore the baby itself, have comes from someone who lives in that world. It is like learning from a book or via the internet without having any experience of the real thing. However, it's the best this growing mechanism can do under the circumstances: copy from someone who knows. Of course, the mother can only know the world she lives in and that is exactly what the baby is being prepared for.

The foetus learns to put programmes in place that will express an appropriate reaction to specific conditions in the outside world. But since it doesn't actually know this world, and does not have to respond to anything, it is all still potential and virtual. For instance, it will learn the physical reactions it will need to display when confronted with various situations that trigger certain specific types of vibrations, such as love, hate, anger, fear, frustration, happiness, sadness, feeling threatened, joy, etc. The foetus records the various different physical pathways a human uses to express different circumstances. It knows what the different programmes are and that they are needed. It also knows that certain programmes are needed a lot more than others, but it doesn't actually know when those programmes are required, because it doesn't meet the outside conditions that induce the different physical conditions. A foetus that develops inside a mother who lives in constant fear will realise that the most normal status to be in is one of fear, and it will learn to use those physical pathways most often. Therefore, certain learned pathways will be used frequently, others less frequently and still others rarely. It is this kind of information that actually gets laid down in the physical tissues. The foetus will develop a physical system that is geared up to meet the circumstances of the outside environment it will enter into.

What Kind of Programmes Do We Find Within the System of a Fully Developed Foetus?
A fully developed foetus has a record of the complete history of the universe within its tissues. So every cell of our body knows everything that has happened between the Big Bang and now. Once we have reached the human form, the foetus will, in its

further development, display the information of the specific lineage it originates from. Here, all the choices humans have made in their physical development will be laid down in the tissues of the foetus. For example, whether your background is Caucasian, African or Asian; or, which country or area your family comes from. Through specific experiences people hold specific beliefs, which will be expressed in the tissues of the human being in the making. It also makes a difference whether the family is from a rural or city background, the cultural background, whether or not they are sea people or mountain people.

All the choices are expressed in the tissues, right up to the family habits and the traits that can be recognised in the physical structures as well as the mental attitudes of families and areas. Certain shaped noses are said to be typical when you are of Jewish descent. This doesn't mean to say that all Jews have this trait, but that it is a common one stretching way back into the ancestral lineage, and seen far less in other communities. Variations in skin colour occur because of the climatic conditions people have lived in. All differences between the various human races and tribes can be related to adaptations to the circumstances these people have lived in for a very long time. These traits are passed on to the next generation, as they are part of the knowledge about the outside world the foetus is being prepared for. It would not encourage the survival of the human race if in Lapland black people were born, all ready and prepared for hot weather conditions. Hence, if we are born in a palace, our tissues should have the information and the structure that is required to survive those palatial circumstances, which would be very different from the ones we would encounter if we are born in the slums. To the child, neither is better nor worse. Disaster strikes for the child when it is prepared for certain conditions and then finds itself in a completely different world, even if that world is, by certain human standards, considered to be more loving and child-friendly. These children do not know how to

behave in such a world, and may have great difficulty in matching
the internal pathways to the external experiences, because their
learned programmes are mismatched with the outside reality.

So, the history of the universe is being put down in human
tissues, followed by the history of humanity, as it was experienced
by the ancestors who directly preceded the building of this new life.
All this information is recorded in the human DNA, which ensures
that the information will not get lost and will allow updates as and
when important things happen that need to be passed on to the
generations to come.

There is only one other possible way of altering the genetic
information within living cells. The creation story shows us that
for information, waves and frequencies to be recorded, they need
to find their way through the various densities of each layer of
creation. For the human form, this means all the way down from
the plasma through to the blood down into the nervous tissue and
beyond into the seed. Only then will it have altered the human
being right down to its roots. Only then will this information be
recorded in such a way that the human race has access to it, not
just the individual through which it has been recorded. The seed
itself has been altered.

From the creation story we also know that when information
hits matter, it loses some of its power and potential, because energy
is transferred from movement into matter. For information to have
enough power to reach all the way down into the deepest levels
of matter, it generally takes time and repetition. Matter is formed
step by step by transferring bits of energy into matter. The next
wave makes a bit more matter of a similar kind, but a different
consistency (getting denser all the time). It is almost impossible
for one bit of information to reach through several layers of density
and be written up into the seed in one go. You need to imagine
something so phenomenally powerful that it bursts your entire
world out off of its hinges. It happens; but not very often. The story
of St Paul on the road to Damascus seems to show an impressive

impact, but it still isn't powerful enough. On the whole, things develop almost layer by layer, and once it has reached the nervous system stadium, it could be transferred into the seed layer. The main reason for this slow and tedious process of development is that you don't want everything you experience and come across to totally burst your entire world. You need time to evaluate, to digest, to contemplate the impact, before you allow that kind of information to be passed on.

It is similar to the fact that as a parent you do not want, nor do you have the urge, to tell your children everything that has happened to you and how you experienced it. Instead, you give a summary of the most important lessons you have learned that you want to pass on. In addition, we can't remember everything that has happened to us anyway. Why not? Things that may have appeared very important at the time, may not have sunk beyond a certain level in our recording system; our tissues. For any information to be included in the seed, it needs to be experienced and recorded in all other tissues, including the nervous tissue. When it has fully sunk to the bottom, only then can we be sure that every aspect of the human being needs to know this information and hence it can be included in the encyclopaedia of knowledge, the latest expression of the universal creation, the human DNA. That is why human beings, like all other parts of creation, have only developed over many, many years. It is all about learning, not being thrown off course by simple singular occurrences.

So, filled up with genetic information, completed by the written physical pathways of possible responses to outside stimuli, what is the baby going to now do with this information?

When a baby is born, he/she will gradually encounter the outside circumstances that triggered the programmes that were written into their tissues. Babies will always look at how their mothers react to circumstances, noises, voices, food, people, animals, images, and the way their mothers respond, the babies will respond, and make the connection to the physical pathways they require in order to express that particular sensation or feeling. Initially, all the choices babies make in reaction to the impulses they receive from outside will be copies of the choices their mothers (or other most important person from their environment) make. Later on, babies learn to observe other people's reaction patterns and they connect with those too, especially if they are more or less in line with what they have learned.

Once a baby is born and has arrived within the world as we know it, the new-born cannot be viewed as an independent life as such. Not yet, at least. The human baby is still not capable of sustaining life by itself. For all of their essential needs, babies remain completely dependent on their mothers. She is responsible for the safety, warmth, cleanliness and feeding of the baby, and thus for the health of the baby. Others can take care of some of the practical tasks for a while, but the truth within nature is that the new-born focuses all of their attention on the mother (in the early stages), as that has been the source of all information during their intra-uterine growth, and is therefore the most reliable source of information for them.

It would be correct to view the life of a new-born baby as still developing within the energy field of the mother, which is not really any different from before birth, but with a different prospect towards the future. This means that the baby takes on board the information from their environment via the mother. The baby will mostly focus their reactions on the reactions of their mother. When the mother is anxious, angry, or happy, for example, the baby lives out her innermost feelings in their own life, using the physical pathways that were written down as a foetus. The baby will express in his/her tissues what the mother displays in her energy field. This also means that the illness of the baby will be a reflection of the state in which the mother's inner life is moving at that time and the conflicts that are present in the baby's environment. If the mother does not feel free to express her fears and anxieties, the baby gets confused; he/she knows how the mother is feeling but they do not experience the expressions that match those feelings. The baby has not yet learned to be that deceitful and can only display the inner truth as they read it. This means that the baby displays the illness, which is always a conflict that the tissues of the mother, at a deeper, hidden level, are also displaying.

Babies continues to record everything they come across in pretty much the same way as they did during their foetal life, by allowing the information to penetrate into their tissues. This time, the information comes from direct contact with the outside world, and the deepest possible level at which information can get stored will now be the nervous tissue. First, babies check out how their mothers connect the outside stimuli to the inner expressions, but gradually, more and more, babies have their own direct connection with a growing number of sources of information. Other people, such as the father, grandparents, peers, neighbours, friends, etc., influence them more directly, and no longer only through their mothers' ways of experiencing the outer world. Babies take on board everything they come across. All experiences are written down in the tissues of the new human being.

When the nervous tissues have recorded the information, you can be sure that every fibre of the baby "knows". The nervous tissues are the densest of all human tissues, with the exception of the seed itself. Electroencephalographic studies (EEG) on babies and young children have revealed that the main brain function during the first two years of life consists of delta waves. These are waves – vibrations – with a frequency between 0.5 and 4 Hertz. The brain activity that a young child displays is one that is seen in adults in a state of deep unconsciousness, or, in other words, a deep comatose state. These are conditions in which the brain only receives information and does not send out very much. We can therefore summarise that babies and toddlers only record experiences, and they do this without any evaluation whatsoever, because "judging" the experiences would require a higher brain function with different vibration frequencies.

Between the ages of two and six, babies and toddlers show mainly theta waves (48 Hz), which corresponds to a lighter comatose state, while still being unconscious. It is considered a state in which nothing else happens but the recording of experiences without asking any questions. In other words, everything gets recorded the way it presents itself. Between the ages of six and 12 we find predominantly alpha waves (8–12 Hz), and above the age of 12 it is beta waves (12–35 Hz). The alpha waves represent calm consciousness and the beta waves happen in a state of active, focussed consciousness. Above 35 Hz, these brainwaves are known as gamma waves and represent peak performance states.

The conclusion is that up until the age of about six, the child is mainly occupied with recording experiences as they happen. We learn to link our outside circumstances with inner functioning. The pathways of what the tissues do when we are happy, sad, angry and so on, are now being linked to the things that make us happy, sad or angry. It is about recording the things we do or don't like, and the reason we do or don't like them is because our environment (first mother and then others) do or don't like them. Each experience is given an evaluation tag by how our environment reacts to the experience. This means that in order to know how to respond in the future to a similar event, each record gets a tag that defines the experience as "pleasant", "worrisome", "dangerous", "hurtful", "painful", "comforting", etc. It's not the baby who does the evaluating; he/she *feels* what physical pathway the mother uses and codes it in the way the mother does. Every single experience gets a code in that way – an emotional evaluation – which explains why we react so violently and so consistently to our environment, and why we are so adamant that everybody "must" feel the same way. This is our unconscious programming telling us "how it is".

So a baby's life is about *recording experiences*. At the beginning of a new life in a new and strange world, the baby benefits by learning as much as he/she possibly can in the shortest time possible about the world they will need to survive in. Of course, that world is different for all of us, and is different for siblings born at different times. All the baby needs to know refers to his/her particular circumstances. The recorded information is written down within the tissues of the baby and will serve as a template for all future experiences. They will be matched with the stored information and this might possibly cause internal conflict, but at the baby stage, all that matters is that information gets recorded. To a baby, life is as it is; they have no connection with an adult interpretation of a "hard" life or an "unfair" life that adults judge some babies to have, because as adults, we live with judgements about life and its circumstances. We think we would like it to be this way or that way, and we dislike it when it is not. We attach blame for what we judge to be unfair in life. Very young children don't do this; to them, life is as it is. It is what it is!

As children grow up, they start to take notice of the effects life and everything in it has upon them. They start to recognise pain as something that hurts and that is quite different from joy, which feels pleasant. Now they start to "like" and "dislike" certain things that happen to them. Now they start to evaluate.

They have their pre-programmed pathways and reaction settings and they use them to interact with the world around them. This interaction has an effect, and this effect is now consciously felt and recorded in a similar fashion. To start off with, the evaluation system is crude and tends to use generalised tags such as "good" and "bad", just as we considered the three basic aspects earlier on: *Hot–cold, Dry–wet, Heavy–light.* Children like feeling good; they don't like feeling bad. Now they will begin to pursue the activities and states that create this good feeling. They make choices about which pathways to employ more often and which ones to avoid. This sets them apart from the people who gave them the information in the first place; their parents, family and environment, which might include church, school, youth organisation, etc. In effect, they try out what they have been given in order to test and evaluate the value of that information for themselves.

All learning is about experiencing things. Humans need to try out all options before they truthfully know reality. We know that human beings will first try out all possible treatments that don't work, before finally settling on the one that does work. This is an essential part of life, because it is only through experimentation that we can truly feel what the result is. We can learn from books and from listening to wise tales, but this won't be real until we have felt it for ourselves. Humanity as a whole needs to make all of the mistakes before it can truly know. It is only through the evaluation system that looks back at how we felt about having a certain experience, that we learn to avoid the ones we don't like and chase after the ones we do like.

The next stage is for us to pursue pleasant experiences, and to try to keep life on the "pleasant" track, just to find out whether that is possible. Not only is it impossible, because life doesn't evolve around the logic of the small human brain, but also because we learn that what we liked before, we might not like anymore when tried again in the future. If we like sunshine and warm temperatures, we might believe that if we moved to somewhere warm and sunny we would be eternally happy. This could turn out to be a fallacy, especially if you come from a family with its roots in changeable climactic conditions. What was once new and exciting could soon become stale and boring. In other words, we start to notice that our evaluation also evolves, it never stands still or remains the same. Hence, we hear people in desperation cry, *"but You Used to Like It!"* Eventually, when we keep pursuing happiness by chasing the circumstances that make us feel good, we come to realise that the feeling good bit is something that comes from within us. We create us feeling good. And we tend to use outside circumstances in order for us to trigger the particular programme that creates "good". The realisation that feeling good is simply our own creation liberates us from the circumstances that are needed to do this.

You might ask yourself: *"When the way I feel is simply a creation of my own inner system, then surely I must be able to induce that creative process irrespective of any possible outside triggers?"* Indeed, being happy, and all other sensations I might produce as a human being, is a direct result of the programmes I set in motion. When I have grown up to the extent that I can make my own decisions independently of my outside world – when it doesn't really matter what anybody else thinks – I can decide to be happy under all circumstances. Most people still find that concept too far-fetched. They will argue that: *"When a Loved One Dies Surely You Cannot Be Happy? Never!"* As a matter of fact, you can! *"Oh, That Means You Don't Care!"* But is being miserable a sign of caring for someone? This kind of argument loses its emotional power – and that is its only power – when you put it in more neutral terms.

I am happy because the sun shines and because it is warm. Now it starts to rain and it is cold, which makes me unhappy and miserable. Or I can decide that I am just as happy in the rain as I am in the sunshine. Why would I do that? Well, do I feel better when I am miserable? Do I function better when I am miserable? Do I gain from being miserable? That is one set of considerations. Another is: Does it stop raining because I am miserable about the rain? In other words, observing this experience I learn that the outside environment does not change just because I feel bad about it. Does my inner environment change from good to bad as a result of it starting to rain? Not really! My inner environment changes *Because I Change.* Because I am the one who "controls" my inner environment.

People still say that there is nothing they can do about the way they feel. *"I Just Feel That Way; I Can't Help That!"* This is because these people have not yet discovered that the trigger is not the result. The result, in this case the way you feel, is caused by the programme that your system is running. The trigger is on the outside; the programme is on the inside. The thing that connects the two is *You.* You are the one who experiences the outside trigger through the filters you have put in place. You are therefore also the owner of the programmes. You decide which recording you are going to listen to. The two things – triggers and feelings – are not one and the same thing; they are two separate things. And we can decide to keep them separate. We learn this after feeling really bad for a long time or feeling bad with a great intensity. Realising that you don't have the power to change the world or the universe might help you to focus on what powers you **do** have and to start using them wherever you can. These personal powers lie inside of you. You don't change the weather or the passing of a loved one; you change your reaction to it.

This is an important stage in the growing up of the individual, as well as the maturing of humanity as a whole. We need to become more aware of our personal power and use it in the field where it is the most powerful – internally. Once children have become aware of their personal power, the "terrible teens" have arrived. Now the growing child is going to try out their newly gained knowledge; they are going to push the boundaries.

Let's now look at this development once again, but this time following the frequencies and information flow as it starts to manifest itself in the tissues and the growing and maturing human being.

Chapter 7
A Human Being Created from Information

The information that flows out of the initial disturbance of the fabric of the universe, and that creates matter and movement, can also be traced to within the development of the human being itself. We now know that there are seven distinct layers within creation that are repeated over and over again. The information that is contained within each layer is the same each time it is repeated, but it is expressed in more and more dense ways until it materialises in seven different layers of matter, which will then once again differentiate into seven times seven times seven varieties.

The initial seed, the third frequency, which is brimmed with *emotion*, containing inner elation and peace as well as explosive anger, starts the evolutionary process by bursting open, allowing its expansive energy (yang) to permeate its surroundings. Yin and yang are images used to express the constantly transforming interactions. Originally, the Chinese characters named as yin represented the moon energy and yang represented the sun energy. Gradually, these terms were broadened to include yin as night and yang as day, yin as winter and yang as summer, and yin as female and yang as male. Yin is that which maintains and endures. It is nourishing and supports growth and development as well as being something contracting and moving inwardly. Yang is that which is creative and generating. It develops and expands. It is dynamic and full of movement. In general, anything that is moving, ascending, bright, progressing and hyperactive pertains to yang. The characteristics of stillness, descending, darkness, degeneration and hypoactive pertain to yin.

Ancient knowledge teaches us that the human body is governed by seven major energy centres, called chakras. Their positioning on the body is in accordance with the Golden Ratio, and each centre is said to represent specific energies, which will create and operate certain parts of the physical body. This means that disturbances within a specific chakra – a specific information field – will influence the functioning of specific organs and parts of the physical body. Of course, the information is distributed to all tissues, but it has a higher prevalence in certain ones, depending on the percentages of contribution of each chakra to all the different parts of the system. These chakras are energy condensations that will then condense even further into matter. The physical form, the quality of this form and the way it will function, depends on the purity and balance of the energy within the chakras.

Let's take another look at this development, but this time by following the frequencies and the stream of information as they manifest in the tissues and in the growing human being that is maturing all the time. The physical form, the quality of this form and the way the tissues work depends on the purity and the balance of the energy flowing through the chakra gates.

First let's look at what can be said about these chakras. What can we learn about the meaning and the function of these gateways between the energetic field of the individual and of his/her physical expression of this field, in particular the body?

- **Chakra 1:** The basic principle of this chakra is *"All Is One"*. The message is one of connectedness with all of life, and for the human being this means that the individual is part of the group, the family. From this a new life emerges and develops.

- **Chakra 2:** The basic principle of this chakra is *"Respect for One Another"*. Here, the way interaction happens between all life forms is determined. This relates to the movement between the inner world of the individual and the outer world. This is where independence begins.

- **Chakra 3:** The basic principle of this chakra is *"Respect for Oneself"*. Here, the emphasis lies on the inner power of the individual, functioning independently within the framework of nature and the tradition of the human tribe.

- **Chakra 4:** The basic principle of this chakra is *"Love Is Divine Power"*. Here, unconditional love is meant, involving an endless capacity for forgiveness. Learning what this kind of love means brings stability into our life by finding eternal peace amongst the love that attracts and the hate that repulses.

- **Chakra 5:** This is the chakra of *"Choice"*. Here, the emphasis is on the choices we make and the acceptance of the consequences. Expressing and communicating those choices, our thoughts and our belief systems confronts us with the reactions of our environment.

- **Chakra 6:** The basic principle of this chakra is *"Only Search for Truth"*. The mind is encouraged to discover what is true and what is illusion. This is the way the consciousness of the individual develops.

- **Chakra 7:** The basic principle of this chakra is *"Live in the Moment"*. The emphasis here lies on the connection with the divine. This knowledge and wisdom encourages us to see things as they really are, and to release our fears with respect to the physical world.

The energy field contains frequencies. These frequencies have meaning and carry information, which is expressed through the filters, which are the chakras, into matter. Chakras are the most condensed parts of the energy field, just as the cell is the most condensed part of the primary tissues. There is no more compact way to organize the tissues than they are in the cell, and now the inflowing energy stream of information gets stuck in the matter, it becomes fully absorbed. The only way to avoid an explosive situation and destruction of this organized matter is to allow the primary tissues to "grow" under the ever-increasing pressure. This happens in the same way that a balloon expands, by occupying more space and volume as the air pressure increases inside.

The continuation of inflowing energy into the primary tissues increases the pressure inside those tissues, and the gateways through which this energy flow enters the matter are called the chakras. *Each chakra has a frequency code and therefore a certain Content*, a meaning, to add to the "airflow". The specifics of this contribution of meaning varies according to the combination possibilities we discussed previously, whilst proportionally splitting up the visible light into the seven colours. This leads to a variety of manifestations and, subsequently, a variety of life forms, as mentioned earlier. The contribution percentages were also discussed earlier.

The development of the human being, code 7 of the creation, starts to happen at a certain point within the story of creation. We are in solid matter, the element earth, and we happen after the plants, the egg-laying animals and the mammals. This means that the basic developments that happened on those levels are embedded within the tissues and structures that human matter forms. So, we can deduce that the energy of frequency 5 will form the sensory system within the human being. Frequency 2 will be the movement system; frequency 6 the breathing and digestive system; frequency 4 the circulatory system; and frequency 1 the lymphatic system. From the place frequency 7 occupies within the creation code, we can deduce that the energetic information will be responsible for the formation of the nervous system. Frequency 3 is the seed tissue with energetic information about personal power and respect for the self. In the human being it becomes obvious that that kind of power is expressed by the excretion system, which keeps the organism clean and functioning. The physical structures that are essential to this are the glands.

The human cell in which energy flow gets stuck, and which is going to grow into a human body, consists, just like any living cell, of the same primary tissues formed out of the same energetic information; one layer emerging from the previous one and so on.

- 39.56% is water   frequency 1
- 24.45% is blood   frequency 4
- 15.11% is muscle   frequency 6
- 9.34% is fat   frequency 2
- 5.77% is bone   frequency 5
- 3.57% is nerve   frequency 7
- 2.20% is seed   frequency 3

The physical composition of primary tissue is the same for all living cells. However, the energetic blueprint of a plant cell, an animal cell, a mammal cell and a human being cell is totally different, because the energy field in which the cell emerges has a totally different composition (for example, a plant is 6–4–2–1–5–7–3). This manifests itself clearly when the original cell grows into an organism. With technology we can establish the differences between the cells on a microscopic level within the intracellular structure, and these differences become the various creatures that form in the macroscopic world – plant, animal, human.

Therefore, the tissues that are used in material creation are very specific, and contain very specific information that is ultimately stored within the DNA of the cell. In order to "protect" these tissues from becoming encrypted with non-relevant information, gateways are erected in every energetic layer of the structure to make a "selection" of the information that is allowed through. Because all energy is polarized, these gateways can keep out energy waves that are not meant for that particular tissue, which results in a selective interference pattern within the cell. This enables the cell to grow in a very specific way, which creates an identifiable organism made up of specific cells.

The chakras contain an individualized set of patterns – thinking and feeling patterns – which the individual has received from the history of mankind, from knowledge of the region, of family experiences and from the parental views of life. This is a filter that scrutinizes the outside energies so as to pick out the specific polarized ones. This means that the way the chakras are composed determines *how* that individual experiences the world around himself/herself, and how to evaluate, judge and respond to it.

It is clear that the energy flowing through the chakras results in a unique structure, which is a sort of magnification of the cell. This structure then functions in a very specific, unique way, guided by the filtered information that enters the formed structures. Neither the structure itself nor the functioning of the structure is determined by the surrounding energies, but is determined by the composition and polarization of the gateways.

Broadly speaking, the physical structures in relation to the frequencies will look like this for a human being:

- frequency 1 – lymphatic system
- frequency 4 – circulation system
- frequency 6 – breathing and digestive system
- frequency 2 – movement system
- frequency 5 – sensory system
- frequency 7 – nervous system
- frequency 3 – excretion and glandular system

Frequency 1 is through chakra 1 and is responsible for the formation of the physical lymphatic system. In addition, chakra 1 is responsible for the formation of water in each of the seven physical structures of the body, because each frequency also shows up directly within the primary tissues that are a part of all physical structures. Each frequency then divides according to its own specific code, which allows us to see how the same seven frequencies form, in different combinations, the tissues.

Let's specify a few things a bit more clearly. **Frequency 1 and 4** manifest in seven sub-frequencies, which are references to the primary tissues; the basic building blocks of all structures. **Frequency 6**, on the other hand, shows two different physical substructures, both of which are manifestations of the split we found in the energy expression of frequency 6: the digestive system has code 6–4–2–1–5–3–7 and the breathing system has code 6–4–2–1–5–7–3. Both are physical manifestations of chakra 6, in which the seven primary tissues are present. From here on, the further development stages, frequency (chakra) 2 and 5 and now 7, will all have incorporated this duality into all of their substructures, because frequency 6, present in all of them, creates a "top" and "bottom" part in all of them. We could say that frequency 6 is responsible for the elongation of what was previously a round shape. Cells become rod-like and the plant grows, so to speak, *Out* of the soil in a direction that is perpendicular to the plane the already developed systems occur in.

The next layer of development then incorporates this "outgrow" within its structure. So, the new structure of the plant – roots and plant – emerges from the existing structure; earth. When **frequency 2** develops, the egg-laying animals integrate the frequency 6 structure inside the body by forming the digestive system – roots – and the breathing system – plant. The new development due to frequency 2 is the movement system; joints and muscles providing more mobility to the animal. First, this happens on the outside of the animal as the exogenous skeleton; the bony suit of armour we find in mussels, and then later as in the crab and lobster. At the next level, it transfers to the inside and the first manifestations are seen in frequencies that have already developed fully, which in frequency 2 development is 1, 4 and 6. When we look at the skeleton of a fish, we see that it consists of a tail (frequency 1), a ribcage (frequency 4) and a neck, of which the head is a part but is not independent (frequency 6).

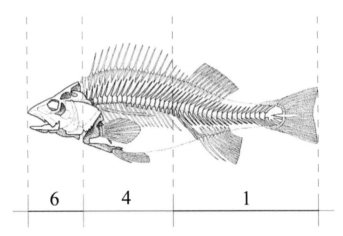

Frequency 2 is about the development of the motor system, and we can follow its growth by looking at the formation of the skeleton. In the skeleton of the lizard (see below), we find the known structures that we have identified as manifestations of frequencies 1, 4 and 6. In addition, we have the development of the pelvis with the hind legs and of the shoulders with the front legs. As frequency 2 develops more and more, we see the refining of these structures and we can deduct that the pelvis and back legs are direct manifestations of frequency 2 within the motor system. The shoulders and front legs also develop during frequency 2, but to start with they do not move independently and these are more an expression of frequency 5, which has the second-highest contribution in the motor system of the vertebrates (code for frequency 2). At this stage of the development (frequency 2), these structures are clearly present, but the independent function will only come at the next level, and in the meantime they are simply supporting the hind-leg function.

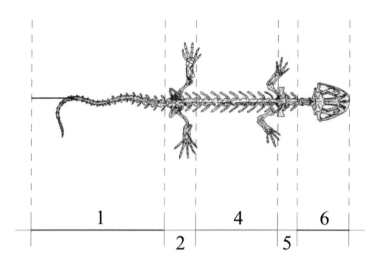

When fully developed, we can link every frequency via chakra 2 to a substructure of the skeleton: 1 tail and sacrum; 2 pelvis and legs; 3 lumbar back; 4 dorsal back and ribs; 5 shoulders and arms; 6 neck and face; 7 skull. Frequency 3 is a rudimentary stage and frequency 7 has only started development into the human being. This will, once again, happen perpendicular to the existing spinal column. Each of these substructures has a "lighter" and a "heavier" part, which relates back to frequency 6. In a nutshell, the substructures of **frequency 2, the motor system,** look like this:

Seven substructures that manifest layer after layer: first **1, tail and sacrum,** followed by **4, ribs,** and then **6, neck.** These substructures are present in the human being in simple form after they have played very important roles in various animal species.

Then appears **2, pelvis and legs,** followed by **5, shoulders and arms.** These substructures occupy a prominent place in the human skeleton as they were developed fully in the previous layer of the mammals, whilst functionally it will still alter slightly in ages to come.

And then it follows to **7, the skull,** and **3, the lumbar spine.** These are not fully developed yet because frequency 3 has not been manifested in its details and frequency 7 has only just begun the journey.

Now, we have to give consideration to the fact that after the initial development within such a layer, the newly formed structure needs to be integrated within the next basic organism structure. Throughout frequency 2 development, we notice that mammals gradually pull the pelvis and legs in line with the spine, as opposed to the lizard where they are away from the spine. This gradually allows mammals to stand up right and use their front legs independently from the back legs; they can grab, dig, fist fight and so on. At the next level, mankind, the upright position is established, and the legs are almost an extension of the spinal column. Furthermore, in mammals (frequency 5), we see that the shoulders and arms grow to have independent functions from the hind legs. At the same time, sensory development (**frequency 5**) grows in an outwardly direction. Arms relate to feeling and touching, but we also see large ears and long noses. In humans these become more integrated in a flatter face and the arms rest alongside the body, in line with the body.

FREQUENCY 5, SENSORY SYSTEM:
First frequency 1 develops, the skin – *the sense of feeling.*
Then 4, which is the fire element that spreads light – *the sense of sight.*
Then 6, the consciousness or intuition – *the sixth sense.*
(These first three are present in plants too.)
Then 2, movement creates sounds by friction – *the sense of hearing.*
(This development starts within egg-laying animals.)
Then 5, subtle communication – *the sense of smell.*
(This is an important part of what mammals are.)
Then 7, collecting knowledge (baby mouth) – *the sense of taste.*
(This begins in mammals and develops fully in mankind.)
Then 3, the seed of information transfer – *the sense of nursing.*

Isn't this amazing! Each creation stage has taken care of a more

refined way of connecting to the outside world, which has manifested itself in an ever-increasing number of physical characteristics
and organisms. The plants, for instance, are aware of touch (1),
recognise light (4), and react intuitively to the circumstances
that surround them (6). They do not have a developed hearing,
smelling or tasting organ. The egg-laying animals add to this a
sharp hearing (2) (yin – incoming, receiving), even fish, and they
use sounds (yang – outgoing) to communicate with one another
and/or to get to know their world, such as the bat using sound to
navigate. Mammals use their sense of smell to communicate with
one another and to learn about their world in greater detail. In
mankind, for the first time the sense of taste plays a major part. If
you doubt this, then have a look at the value that is given to taste
in food adverts and the time the media spends on recipes and
cooking programmes. In Ayurveda, the healing properties of food
items and herbs is determined by their taste.

We know our five senses, and most of us will accept the mysterious sixth sense, even if it is reluctantly, but using seven senses is
surely a bridge too far, no? Well, if we stick to the energetic developments, and the subsequent physical manifestations, it makes
sense that there indeed exists seven senses. Breast development
and the function of it only appears during the development of the
mammals. It is a glandular development, which is a frequency 3
department, and only when the senses develop (frequency 5), does
this form of communication manifest. It is a way of communicating
(frequency 5) between mother and baby with a content of caring,
protection and recognition from mother to baby, and from baby
to mother a need for safety, security and recognition. The result
of this exchange is a force for growth (frequency 3) within the
new-born. We just invented the seventh sense!

Let's fit some examples into this human manifestation scheme.
How can we begin to imagine the relationship between the energetic input (charka) and a specific part of the body?

The human being manifests as **frequency 7** within creation,

which has an energetic code of 7–5–2–6–4–1–3. From the moment
frequency 3 breaks open, a cellular structure becomes evident in
material terms. Seven primary tissues build seven primary systems
within a cell, and later within the developed material structure of
the human being. And we have codes for this process, as follows.

Frequency 3 breaks open into seven layers that form tissues
and systems:

1–3–4–6–2–5–7	lymphatic system	glands: lymph, sebum, sweat
4–6–1–2–3–5–7	circulation system	glands: spleen, kidneys
6–4–2–1–5–3–7	digestive system	glands: liver, pancreas, bile, stomach acid
–7–3	breathing system	glands: thymus
2–5–6–7–4–1–3	motor system	glands: adrenaline
5–7–2–6–4–1–3	sensory system	glands: saliva, tears, milk, ear wax, nose mucus
7–5–2–6–4–1–3	nervous system	glands: pineal, pituitary
3–1–4–6–2–5–7	excretion system	glands: thyroid, gonads

This lists a number of examples of the physical manifestations of
frequency 3 within each of the primary systems. It is by no means
an exhaustive list. Within each primary system, all other systems
are represented to a more or lesser degree dependent on the place
the frequency occupies within the code. When one system splits
up into substructures, as is the case in 2 and 5, then each system is
represented by organs within each substructure.

Each chakra manifests within a *System*, within each *Subsystem*,
and in each *Primary Tissue* from which all other structures are
constructed.

The Making of a Human Being

In the course of our study with regard to the universal code and
the code of creation, we have identified the seven layers of man-

ifestation in which energy transforms into matter. We now know that the whole of creation is an endless repetition of an infinite variety of combinations of the same seven frequencies. We have also identified the human energetic code and the code of the seven tissues from which every cellular organism that exists within the universe, including the human being, has been constructed. Now we will explore the way the human body grows from a foetus into an adult.

The development of the foetus starts with conception. The first magical cell, different from the host cells, together named as the woman, appears. Viewed from codes and frequencies, how can we define a cell? Remember from the code of creation in which day 1 (frequency 1) corresponds with the creation of heaven and earth, whereby energy (waves) and matter (particles) become separate. Day 2 (frequency 4) is the creation of land and water, separating matter into two distinct forms. Day 3 (frequency 6) creates plants. Day 4 (frequency 2) creates egg-laying animals. Day 5 (frequency 5) creates mammals and day 6 (frequency 7) creates the human being. On day 7 (frequency 3) we "rest"; this is a day in the future, it is yet to happen.

The cells that form at the end of frequencies 6, 2, 5 and 7 have one thing in common: frequency 3 is always the smallest, most compact, energy band. The contribution of each of the seven primary tissues is for all material cells the same – percentages of water, blood, muscle, fat, bone, nerves and seed – but the "quality" of these primary tissues differs in the various layers because the energetic fields in which they manifest are completely different each time. In the matter, we recognise that the cells of those layers have a different "focus", a different priority. The cells of the primary tissues are very specific to the species because they contain information about the kind of proteins, sugars, etc. that these organisms require. This means that a cell of a grain of rice produces everything in a combination that is uniquely made to end up with the seed that will produce rice. This same principle can be found in all animal species and also in the human being. Even though the same molecular structures are used by all, it is their combination and the way they function together that is different. Cells uniquely belong to specific species.

Furthermore, when we jump from frequency 7 to frequency 3, producing the first new cell from which a new specimen will grow, we need information from other specimens from the environment to complete the starting cell. Plants create flowers – the procreation organ of the plant, which consists of a female part, the pistil, and the male part, the stamen. The latter produces pollen and disperses this into the air. The pistil collects the pollen so the plant can be fertilised. It receives the pollen from other plants from the same species, not from itself. In this way, the plant communicates with the outside world in which it lives, and with this information it can alter itself according to the changing environment. It mixes information from the surrounding same species of plants into its own complex structure, and incorporates this into its DNA. A similar process happens in the animal kingdom and the human world, whereby conception is the point of mixing information of another specimen of the same species with its own information in order to create a new specimen. No conception takes place when the information of a different species enters an organism. Only when the combination of information sets is completed does a new specimen begin to grow.

Let's have a look at a human seed cell formed within the third frequency of the mother. It has to be the third frequency because this is the universal seed frequency. It is also the frequency from which all tissues are created. Tissues emerge from the breaking open of the third frequency, and the code we identified is 1 (juices/skin), 4 (blood), 6 (muscle), 2 (fat), 5 (bone), 7 (nerves), and 3 (seed). These are first present inside the primary cell; the first cell after conception. From now on the development is driven by the code imprinted in the DNA.

Earlier we made reference to the fact that a direct joining of the male and female DNA was an impossibility because they were not two identical strands. However, scientists have proven that the DNA of all human specimens, independent of race or colour or background, is over 99% identical. All the differences the medical world is playing with are, in fact, less than 1% of the total genetic information encrypted in the DNA. We have already seen that the genetic material is actually needed to create the first recognisable human form. Seen from that angle it would appear then that the DNA from both father and mother are almost identical and that they could well form a double-stranded helix, except for the end bits, so to speak. The bringing together of the genetic material of two people creates a complementary double set of genes. This shows that the coming together of a human strand with a non-human strand can never result in a pregnancy because it would not have the information to develop into a human being. And a final conclusion can be made that genetic defects found in normally developed babies (or even later in life) will be extremely rare, as this stage of the human development contains information of the universe and the distant human past, not about the later growth process of the person.

The initial growing of the first foetal cell is based upon the expression of the DNA information. This tells the story of our background. The seed cell belongs to the human race and it holds information about people and history, and the experiences these people have had, which divides us into the Caucasian, African and Asian races. Inside these groups we have information about the living conditions of the particular tribe, such as the climate, whether they live at the seashore or in mountainous habitats, and so on. Next, there is information about the group itself, such as their belief system, their habits and family issues. This information flows out from the genetic material in vibrations and frequencies. Seven main frequencies deliver the embryo information to the energy field of the third frequency of the mother, which will help to "shape" the emerging energy into a *Fixed Format*. These seven frequencies, as we have seen, contain information about aspects of life and the universe. In the order they appear in the universe, and equally into every embryo, they bring the following information into the matter:

1 – creates form (is the first and least compact energy and matter)
4 – creates balance
6 – brings about consciousness
2 – brings movement
5 – adds communication
7 – gathers knowledge (very compact tissue)
3 – brings personal power (coded tissue)

The primary tissues that are present within the foetal cell will be fed and stimulated by the constant influx of environmental energy, which allows the material form, the cell, to expand into tissues and systems.

Every seed cell, plant or animal, has the same code for the tissues: 1–4–6–2–5–7–3. The information held within frequency 3 is expressed here and the evolution and formation of the organism carries on up to the highest evolutionary layer the DNA contains information about. For plants this is layer 1, layer 4 and layer 6, because they emerge on day 3, which is frequency 6. Egg-laying animals move up one layer to 2. Mammals express up to frequency 5 and human beings up to frequency 7.

When we look at the development of the foetus in its early stages, we notice that it takes 10 to 12 weeks to reach the human form. As mentioned earlier, first it takes on the shape of a ball; a cluster of cells. This changes into a water-filled balloon lined with a thin layer of cells. Then it becomes a bean, followed by a fish, then the shape of an ape before we recognise the human features. These development stages correspond to the days of creation and are therefore a reflection of the development of the entire universe, right up to the introduction of the human being. It is the DNA of the seed cell that is responsible for the emerging of the human form, and it takes 12 weeks to complete the process. If the genetic information is intact and undamaged, the human form appears without any interference from the environment, being the mother. When there is damage, a spontaneous abortion will take place – a miscarriage will happen – because the specimen is only able to survive if and when all aspects of the automatic development are present and correct, which otherwise cannot be fed by the specific species it is supposed to belong to. There are deformations that are compatible with life, but take notice of the fact that these are still recognisable as human beings. Anything else at this stage of the development is not possible. At the age of 12 weeks, the foetus harbours all systems in the correct format.

After 12 weeks, the exuding power of the seed has diminished, all coded energies have been used up, and the foetus will have to look elsewhere for stimulation. Since all interactions in the universe are energetic exchanges, we need to look for a source of new impulses for the foetus. We already know that the foetus exists within the mother and we are familiar with the concept that the mother "feeds" the baby. But how is this done and what kind of information is taken in by the foetus? The foetus has a human shape – it is human – therefore the energies that can enter it, that it can react to, will be human energies.

Humans have been created from frequency 7, which is composed of other frequencies in the order of 7–5–2–6–4–1–3. Both mother and baby have this code for their energetic field. The mother's expression of that code also contains information about her own particular life and her experiences. The foetal energy field, fitted with chakra filters, is surrounded by the mother's specific field, which is coded the same way as its own field. The foetus will now extract information from the mother's field, layer by layer. All energy information is polarised and in that way the foetus absorbs whatever is required for its growth and development.

This information is then stored within the tissues and systems of the foetus. As the foetus already has a basic structure, the newly entering information can underpin the basic structure and enable the foetus to expand the basic structure into a much more specific conglomerate of cells and organs, filled with specific "mother" information. When, on the other hand, the incoming information does not support nor stimulate the growth of the parts of the basic structure, those tissues and structures remain weak, because of a relative lack of flow of energy.

Strong basic structural features that resemble the father can still be fed via the energy field of the mother in so far as the mother enables that energy to flow through her own field. This we can identify as things she admires in her partner, or things she supports or fights against. The input of that information will always get to the foetus via the mother's field and will have added connotations, but in spite of that it can still stimulate elements within the foetal development. This is how the foetus can expand all of his/her systems fed on energetic impulses that reach it via the mother, but that can contain information about the father too. It will store this information and complete its growth until 36 to 40 weeks of pregnancy. When all data has been collected, it is time to leave the classroom. It is time to move from one world into another; it is time to be born.

When the baby enters the world we live in, it is basically, again, a seed. There is a lot of potential but it is still locked away and still needs to develop, just like any other seed. The baby enters the world as a human energy field, with a code of 7–5–2–3–4–1–3, and with all chakra filters in place. Because the whole already has a specific development in its pocket, it is already expressed in a very detailed and particular way. The first basic structure was built upon information stored in the DNA, which allowed the newly formed structure to be placed into the right environment it would later enter. This information basically comes from the universe and contains a summary of everything that is known up until now, including the background environment of the families of the mother and father. This is a very specific structure of a very specific human being. This is then topped up by the inflowing information from the mother, which tells the baby about the immediate situation he/she will enter at birth, and prepares him/her for it. This sets the tone and the expectations for the coming life the baby will be born into. That is the complete product of the baby entering the human world at birth, with chakras that have been fine-tuned to the specific information needed in order to expand and grow the specific structure of this new human being. A new-born baby is by no means a clean sheet!

Now the baby is exposed to the humans of the universe. The energy field of the baby will now be fed by the energy fields of other human beings, which will allow him/her to feed the tissues with the information that is required to make them grow and to reach the potential of the basic uniquely formed structure. Of course, universal energy is also present at all times and has the potential to get some information passed the chakra gateways. Both the human information and the universal information will be able to enter and "feed" the baby when it is relevant to the structure and polarised in the right direction. This information will make the particulars of the individual expand, grow, become more powerful and undeniably recognisable. Insecurity happens when the newly incoming information does not match the tissue-stored information from the pregnancy. This creates tension within the tissue; it can cause the tissue to become weak and to malfunction.

This is the fully structured human physical form as it constructs itself out of energetic information. The end result, in the mental as well as the physical expression, is a combination of information held within the DNA and information from the environment, and this last one comes in two stages: first the mother's environment and then the human field environment. So, following the kind of information the specific human being receives, we can understand why the physical and mental makeup of a person is the way it is. The question now is: *"How is this human being going to live?"*

When we look at how an actual life develops, we can recognise, once again, seven phases in which the baby grows and eventually dies. We said before that the new human field receives information from the human field that has a specific composition, whereby frequency 7 has a lot more influence than frequency 1 or 3. So, our development is also staged through seven layered compartments:

7 – baby	lives out of universal energies; reacts intuitively
5 – toddler	develops communication skills, makes noises, begins to talk
2 – child	examines its environment through movement, sport, dance
6 – teenager	becomes aware about the self, creates an identity
4 – adult	finding balance in life: homework, family values, partner/children
1 – post-menopausal	focus on own life, more value on self
3 – pre-death	only own needs are important, totally self-centred

Let's point out here that during the development of frequency 6, the teenage stage, frequency 3 appears for the first time in its most condensed form, which stands for the development of the glandular system and for self-power. This is why during the teenage stage sexual development starts; the seed development.

And there it is! Life is a development of three important but separate stages, in which each stage finalises a particular part of that development. Three times a seed is planted in a specific environment field in order for it to grow. First, we have the DNA seed contained within the primary cell. This takes the cell up to the human embryonic stage. The second seed is the human form within the environment of the mother's field. This takes it up to the human foetal stage. The third seed is the new-born baby within the environment field of human beings and the direct exposure to the universal field, which, of course, has always been present. This takes it up to the human stage.

Going through these stages gives us the opportunity to experience specific aspects of life that we absorb through the chakras. These are constantly pressurised by the never-ending stream of incoming information that needs to be given "a place". Each life has a unique absorption filter, and through the reaction pattern on incoming information each life gives back to the human field information. It is in this way that frequency 7 of the universe develops, slowly but steadily. A lot of information, through our experiences, is stored in our tissues and will be released back into the entire field at the time of our death. This changes the information carried within the human field and will therefore change the information that will go into the seeds of the future; the genetic material of the still-to-come human being. In that way the human race can experience the whole range of knowledge without a single individual having to go through all the possible lives that are required to gain that information.

By following the lines set out by the codes, we can begin to see how the various layers of development relate to one another, and we begin to understand the laws of the universe. This allows us to separate truth from misconception and how dumping misconceptions is an excellent start.

Making Babies

They say that *Finding Answers Creates More Questions*, and my own search for answers to the fundamental questions of life is no exception. Once you have found the code for the creation of the universe, you want to explain everything using this code. And surely that needs to happen, but it brings to the surface a whole bunch of questions that emphasise your limited knowledge about the code and how small your consciousness is. Once we have explained how babies grow and humans develop, we can no longer avoid the question of conception. How does that work in terms of energies and material expression? How does information from two separate fields get tangled up to create something new?

In the first place, it should amaze us that there are differences between man and woman, which leads to the question, *"Why is that?"* When we look at the frequencies and their compositions again, it remains amazing that frequencies 6 and 2 have more than one possibility. What are they for? To provide more options? Yes, but the universe does not need more options because it controls where it is going anyway. Then let's have a look at expressions of energy to try to find an explanation.

Something that is remarkable is the big difference in the structure of the male and female seeds. The two seed cells, eggs and sperm, don't look like one another at all, in spite of them having the same energetic code. Until now we concentrated on finding similarities, which has led to the discovery of the codes. But maybe now it is time to pay some attention towards the differences.

It is in the plants, frequency 6, that for the first time a complete cell is formed that will ensure procreation of the species. This cell can create a complete specimen. The information necessary for this is locked within the coded DNA of the cell, with all the necessary building ingredients present within the cytoplasm of that cell. We know that frequency 6 leaves us with two possibilities for development, and we have identified these in the physical expression as the root system, downwards development, and the plant system, upwards development. We have also noticed that the flower of the plant, its seed product, contains a duality in the pistil and the stamen. Conception happens via the pistil, which catches pollen that has been released by the stamen of other plants of the same species. From this, a new cell forms, which has the capability of creating a whole new plant. Here we find the female and male principle for the first time.

When we decode frequency 6 as 6–4–2–1–5–3–7, we end the series on frequency 7. However, frequency 3 is the seed frequency, activated when it is the smallest, most compact of the frequencies. So, this particular line of coding does not deliver a structural seed form. It only has DNA matter in the information frequency of 7. This DNA is released into the environment in the form of pollen, and thus becomes available to other plants of the same species. This allows for new information, coming from outside of that one plant, to be incorporated within a new development. This also allows for a fast adoptive behaviour in relation to changing environmental conditions. The actual cell form of the seed is held within the other part of frequency 6, 6–4–2–1–5–7–3, being the pistil of the plant. Here, the captured DNA information is integrated within an "almost cell" to complete the frequency 7 information layer. Now the plant has a new and complete cell, from which it can develop a new specimen.

What about the next layer of creation, frequency 2? Here, the integration of the female and male principle is carried by separate entities altogether. In the invertebrate, as well as in the vertebrate, we find a complete division within the code as well as observing the emergence of male and female specimens. In the code we find a shift between 4 and 7: 2–6–5–4–7–1–3 and 2–6–5–7–4–1–3. Both lines end in frequency 3 and are therefore reproductive cells. In the material expression we now need to look for two very similar-looking cells and organisms that differ in only relatively small details. One has a predominantly 4 complexion and one predominantly 7. What does this mean? One gives more weight to circulation – more heart energy – while the other is more knowledge-oriented – more brain energy. Two cells, and consequently two organisms, in which one relies more heavily on characteristics related to the heart, and the other on characteristics related to thinking. This sounds like classical stereotyping of women and men. This would mean that the female vertebrate will have the code 2–5–6–4–7–1–3 and the male the code 2–5–6–7–4–1–3. And again, in order to create a new specimen, we require information from both male and female specimens and we organise natural ways to bring this information together. As the species evolve, the joining together of this information happens in various ways because the physical structures evolve too, from the lower egg-laying animals to mammals.

We know that the options in frequency 6 remain as a physical manifestation in the further layers of development in the form of the digestive and respiratory systems. The same is true for frequency 2 because two possibilities remain in this case in two separate entities, being a male and female. The difference is in the development and expression of frequency 2, which creates the motor system. Let's have a closer look at the code for 2 in humans: 2–5–6–7(4)–4(7)–1–3. From this, we could deduce that minor changes should then occur in the structure of the ribcage (expression of 4) and the skull (expression of 7). The major differences should be visible in frequency 2 itself and that, as we have seen, is represented in the motor system as the pelvis. Indeed, when we compare the male and female pelvis, it is clear that the shape of the female pelvis is significantly different from the male pelvis.

The cellular structures of the human seed, frequency 3, differ between male and female in content, as different energetic information is locked up within it. The genetic material within the female egg cell is surrounded by a lot of glycolipids and other nutritional elements (more yin). The male cell is much simpler and contains relatively more genetic material in comparison to other structural content, as it is virtually void of nutrition (more yang).

The difference between man and woman lies in the amount of input between frequencies 4 and 7 within that first cell. More 4 means a female specimen, more 7 a male one. Looking at the development of the physical bodies, we notice that the gonads are formed from frequency 3, which leaves the female with a greater input of circulation (frequency 4 – menstruation) and the male with greater influence of the nervous system (frequency 7 – "*a Man Thinks with His Penis*"). Another specific feature is the development of breasts in the female, and we have placed that within the sensory system as an expression of frequency 3 in 5. Now we see how it fits! When does the development of the breasts start? At the beginning of puberty, frequency 6, in which frequency 4 plays an important role (second place in the code), which stimulates the female parts in all structures, including in frequency 3. When we move into adulthood then the breast tissue is fully functional and adulthood *is* frequency 4 of our lifecycle. Breast tissue only develops in the female, as in the male, in frequency 4, and has a low input in frequency 3, which then does not develop very much in the sensory system, frequency 5.

Let's move on to the question of how these carriers of two seeds of necessary information are brought together.

A man is attracted to a woman and vice-versa. Mutual attraction allows those two fields to close the gap between them and, in ideal situations, they merge. At this point we are not concerned with the quality of this merger and the possible consequences for what will be created in various possible scenarios of the merger. Here, we focus on how the first embryonic cell gets created in relation to the energetic merger. The three lowest chakras are polarised yin in the female and yang in the male (explained later), which means that the merger will happen in a movement away from the male towards the female. The female receives. Now she harbours the heaviest seed information (yin), filled with high potential and nourishment, whilst the male seed has less nourishment and almost pure genetic material; it is lighter, more yang. The merger of those two expressions, two opposite poles, of frequency 3 "lights up" the nutritional soup, and brings life into the potential cell.

What determines the sex of the foetus? This human soup creates a human cell, with the energetic code 7–5–2–6–4–1–3, with the seven primary systems. In frequency 3, the glandular system, which includes the sexual orientation and the gonads, the contributions of 4 and 7 determine the sex. We have already shown that the coming together of the two gene strands of mother and father is not a problem. It will only be the interaction, and the determination, between 4 and 7 that will put them in their respective places, which then is the only difference between the two possible seed cells for the development of the embryo. This is the first cell of a new human specimen, and is complete and inclusive of the sex the fully grown organism will be. This is already locked into the DNA of that first cell in the way that frequency 3 is oriented, and from here on in it is either male or female and cannot be changed anymore, with frequency 4 either following or preceding frequency 7.

The actual contribution of every material building block that will be used to construct the entire organism is included in that first cell, and is determined by the energy field that now holds the cell at its centre. From this basic cell structure, all primary structures will grow, as determined by the energy field of the cell, which will be visible in the completed organism itself. The human cell grows into a human structure with human chakra filters. During its construction it runs through the stages of development that the entire universe has gone through right up to this point. This becomes visible as the various physical structures emerge stage by stage. This explains why the structure of the plant remains a feature of the human structure in the expression of the digestive and respiratory systems. In the same way, the essence of the egg-laying animals remains present in the motor system (muscle and skeleton) and in the sexual differences between male and female. The essence of the mammals remains in the presentation of our sensory system, and in the same way our nervous system will develop. It will take time and the few generations we encounter in our lifetime don't seem to show any structural evolution in that respect, but many thousands of years will. These basic differences are already present within the human foetus when it has reached the stage of completing the human shape after 10 or 12 weeks of gestation.

Once we have reached this stage of foetal development, the growing entity will be washed over by human energies, and nurtured by the human field it finds itself in; a mother with the human frequency code of 7–5–2–6–4–1–3.

Now we know how the genetic information of mother and father are joined to form the first embryonic cell. From the previous explanation, we already know how this cell grows into a baby and after birth becomes an adult human being. This is how humans are created over and over again.

We have just unravelled another mystery of life.

Chapter 8
The Yin–Yang Rhythm of Life

The representations of yin and yang are indicative of the opposing directions in which the energies within life fluctuate: either in- or outwardly. In each of us there is always a balance, although this balance obviously varies with time and circumstances. The balance is also very personal, and the changing of this balance within each of us is very personal. There is no set time for all of us to move from one phase to another. An age-band can be identified in which we expect *Most* humans to make the transformation. These times are traditionally marked by initiation rites that symbolise the growing up of the child. The individual child is, of course, different from all other children. However, we can group them, as we did when we talked about the seven different types of human beings.

Male–female

This idea of grouping in order for us to categorise and to be able to see patterns is important, not in nature, but within the small confines of the human brain so we give "order" to things. A first and very simple way of grouping humans is by the differences we observe between men and women. In our modern Western society,

we are encouraged to smooth out these differences and to pretend
that men and women are all the same; they have the same "rights".
And we interpret this to mean that they have the same jobs to do
in life and that they carry responsibilities for the same aspects of
life. Natural differences show us something else though. We can fill
in the fundamental differences between men and women, because
these manifest as a result of the distribution of the yin–yang aspects
of energy, or put better, the polarisation of energies. Quantum
scientists talk about these in terms of right or left "spin".

Knowing the order in which the chakras appear allows us to
determine the yin–yang direction of each chakra. We know that
the visible light spectrum is divided into 39.55906% red, 9.33863%
orange, 2.20455% yellow, 24.4488% green, 5.77156% blue, 15.11025%
indigo and 3.56707% violet. This gives us the colour proportions
according to the various codes for each colour.

We know that each colour contains all colours, which has
allowed us to figure out how each chakra is composed. Suppose
each chakra has a 100% value. This will be divided into 39.55906%
+ 24.44888% + 15.11025% + 9.33863% + 5.77156% + 3.56707% +
2.20455%. This distribution will be the same for all the chakras, and
only the colours that fill in the various percentages will differ.

- From the code for chakra 7 (7–5–2–6–4–1–3), we can conclude
  that chakra 7 will consist for 39.55906% of frequency 7 (nervous tissue), for 24.44888% of frequency 5 (bone tissue), for
  15.11025% of frequency 2 (fat tissue), for 9.33863% of frequency
  6 (muscle tissue), for 5.77156% of frequency 4 (blood tissue),
  for 3.56707% of frequency 1 (plasma tissue) and for 2.20455%
  of frequency 3 (seed tissue).
- From the code for chakra 5 (5–7–2–6–4–1–3), we can conclude
  that chakra 5 will consist for 39.55906% of frequency 5 (bone
  tissue), for 24.44888% of frequency 7 (nervous tissue), for
  15.11025% of frequency 2 (fat tissue), for 9.33863% of frequency
  6 (muscle tissue), for 5.77156% of frequency 4 (blood tissue),

for 3.56707% of frequency 1 (plasma tissue) and for 2.20455% of frequency 3 (seed tissue).

- The same goes for all other chakras.

We now know the composition of the various chakras, which allows us to calculate the polarisation (yin–yang) per frequency for each of the chakras. Let us start with chakra 7 to see how the corresponding energy field is composed.

Chakra 7: VIOLET 7–5–2–6–4–1–3

An energy field is always looking for balance, for harmony, for resonance. If we make violet 7 yin then the natural tendency is to attract a yang value, and with the next appearing energy being blue 5, this one will then present as yang. Going through the code in this way we can then add a yin (-) or yang (+) value to each figure. The - and + signs are simple conventions and have no "better" or "worse" value.

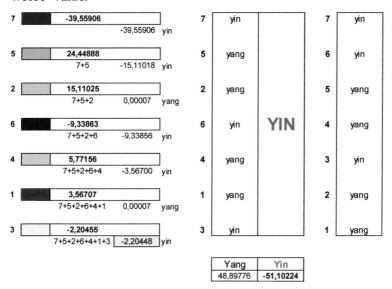

The sum of 7 + 5 or -39.55906 + (+24.44888) equals -15.11018; still negative and therefore a yin value. This means that it will further attract a yang value, and therefore 2 must also be yang in order to move closer to the harmony point, which has a value of zero.

7 + 5 + 2 or -15.11018 + (+15.11025) equals +0.00007; a yang value.

This will now be followed by a yin value, which means that the next number, 6, becomes yin to bring the momentum back in the direction of the balance point. Add -9.33863 to +0.00007 and the result is -9.33856; a yin value.

This is then followed by a yang value delivered by 4. Add the value of 4, which is +5.77156 to -9.33856 and the sum is -3.56700; still a yin value. This means that the next, which is 1, must be yang.

(7 + 5 + 2 + 6 + 4) + 1 or -3.56700 + (+3.56707) equals +0.00007, and is therefore yang.

(7 + 5 + 2 + 6 + 4 + 1) + 3 or +0.00007 plus -2.20455 equals -2.20448, which means that chakra 7, in total, indeed gets a yin value.

According to the code 7–6–5–4–3–2–1, this becomes for a yin chakra 7:

7 yin–6 yin–5 yang–4 yang–3 yin–2 yang–1 yang

In a similar way, we can determine the distribution for a yang chakra 7:

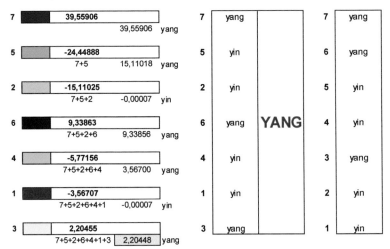

The yang chakra 7 becomes:

> 7 yang–6 yang–5 yin–4 yin–3 yang–2 yin–1 yin

Chakra 5: BLUE 5–7–2–6–4–1–3

If we do the same calculations for code 5, we will find:
Yin–value of chakra 5 becomes:

> 7 yang–6 yin–5 yin–4 yang–3 yin–2 yang–1 yang

Yang–value of chakra 5 becomes:

> 7 yin–6 yang–5 yang–4 yin–3 yang–2 yin–1 yin

Chakra 2: ORANGE D 2–5–6–7–4–1–3

If we make the same calculations for code 2 D, we will find:
Yin–value of chakra 2 code D becomes:

> 7 yang–6 yang–5 yang–4 yin–3 yin–2 yin–1 yang

Yang–value of chakra 2 code D becomes:

> 7 yin–6 yin–5 yin–4 yang–3 yang–2 yang–1 yin

Chakra 6: INDIGO B 6–4–2–1–5–7–3

If we make the same calculations for code 6 B, we will find:
Yin–value of chakra 6 code B becomes:

> 7 yin–6 yin–5 yang–4 yang–3 yang–2 yang–1 yin

Yang–value of chakra 6 code B becomes:

> 7 yang–6 yang–5 yin–4 yin–3 yin–2 yin–1 yang

Chakra 4: GREEN 4–6–1–2–3–5–7

If we make the same calculations for code 4, we will find:
Yin–value of chakra 4 becomes:

7 yin–6 yang–5 yang–4 yin–3 yang–2 yin–1 yang

Yang–value of chakra 4 becomes:

7 yang–6 yin–5 yin–4 yang–3 yin–2 yang–1 yin

Chakra 1: RED 1–3–4–6–2–5–7

If we make the same calculations for code 1, we will find:
Yin–value of chakra 1 becomes:

7 yin–6 yin–5 yang–4 yang–3 yang–2 yang–1 yin

Yang–value of chakra 1 becomes:

7 yang–6 yang–5 yin–4 yin–3 yin–2 yin–1 yang

Chakra 3: YELLOW 3–1–4–6–2–5–7

If we make the same calculations for code 3, we will find:
Yin–value of chakra 3 becomes:

7 yin–6 yin–5 yang–4 yang–3 yin–2 yang–1 yang

Yang–value of chakra 3 becomes:

7 yang–6 yang–5 yin–4 yin–3 yang–2 yin–1 yin

As we know, frequency 3 in the first cell of the new embryo has already made a choice; it has already formed in either a female or male configuration, which we will indicate by using the signs - and + respectively, then we can deduce what the basic polarisation will be for the energies within the female and male specimen. The first cell comes from code 3′, which will open up into seven different systems, which will build the entire human being. Right at the very beginning, the specific human field will become polarised, even before the tissues are blown up into an organism. This means that the female field of this cell will be - and the male +, or to put it differently, frequency 3 will be - for girls, and + for boys.
Frequency 3′ is then - (girl) yin, developing 7–5–2–6–4–1–3, or

7 yang, 5 yang, 2 yin, 6 yang, 4 yang, 1 yin, and 3, already known to be yin.

What does this mean in terms of the general characteristics of the traditional female?

1. Yin: the form is *Stability* (forms the basis for community and family structure).
2. Yin: movement is *Inwards* (attracts what she needs, including a mate).
3. Yin: the power lies in *Personal Internal strength*.
4. Yang: balance is mainly *Giving Love* (gives all with love in order to receive love).
5. Yang: communicates mainly by *Talking* (all the time, including gossiping).
6. Yang: awareness is *Imagination, Intuition* ("knows" things instinctively).
7. Yang: expresses what she knows through *Doing* ("a woman's work is never done").

And for the traditional male?

1. Yang: the form is *Flexibility* (can take it or leave it; easily accepts changes readily).
2. Yang: movement is *Outwards* (pushes things away, work is in the outside world).
3. Yang: the power lies in *Radiation* (meaning something in the outer world).
4. Yin: balance is about *Taking Love* (buys love; takes it wherever, whenever, however).
5. Yin: communication is about *Listening* (takes it all in, says very little).
6. Yin: awareness through *Insight* (searches to understand and comprehend).
7. Yin: expresses what is known through *Thinking* (rationalises everything).

This means that the top four chakras have an opposite polarity to the three lower ones. Female means that 3 is yin (per definition), as are 2 and 1, but 7, 6, 5 and 4 are all yang. For a male, it is the other way around. This means that a female gives great stability and support (lower 3 chakras) but is furthermore focussed on the outside, and the energies are directed towards something or somebody else all the time. They are also better connected to the "higher" realms of life. A male, on the contrary, is unreliable for basic needs (lower 3 chakras), but gathers knowledge and understanding by observation, expressing itself in science. You could say that the way males and females are made serves the need for relationships. Females provide a stable environment for the family by attracting the energies that build a nest and by spreading the higher energies around. Males, on the other hand, are less concerned with building a nest; they are also more likely to move from nest to nest. Males draw to themselves the higher energies concerning love, awareness and knowledge. This separates the receivers from the various energies and brings the two types of humans together on the basis of needs. The strengths of females are needed by the male to make life easier for him, and vice-versa. Together they are much stronger than the sum of the two elements, provided they are allowed to do what they naturally do best.

The Process of Life

When a child enters this world, it starts a growth and development process that has seven layers, with each layer being specific about certain information aspects of life itself. Life consists of seven phases, which each represent different challenges and possibilities of learning different aspects of life.

> *Baby* – has a higher knowledge, an obvious connection to universal information.
> *Toddler* – is about developing speech and listening to orders.
> *Child* – time to be interested in sport, movement and activity.

Teenager – becoming aware of own personality, of needs of
the self.

Young Adult – balance between giving and taking, between
caring for others and self.

Middle Age – finding stability in life and wanting to develop
self.

Old Age – learning to use internal personal power, both for
stability and for peace.

The baby seems to know so much more than the adults who take
care of it. The baby always seems to do the right thing (doesn't fall
of high ledges), and seems to have the similar deep-rooted instincts
that we observe in wild animals. It just is. It just does. We find
babies, of most animal species, cute and charming, as they have
an innocence we know we have lost and we crave again. So, the
baby phase expresses the fundamental universal knowledge this
creation came into being with.

The toddler seems all about exercising the voice. Making noises,
repeating sounds almost indefinitely, and at the same time being
confronted by firm and loud voices that are trying to get messages
through to them. Life at this stage of human development is about
learning the communication skills of talking and listening.

The child rapidly expands its range of exploration of the surrounding
world. They are all about movement, such as cycling, running,
climbing, playing, and getting interested in sports. The child wants
to learn to move and to know what it means to be able to move. It
is, at this stage, more interested in playing than in schoolwork.

The teenager is becoming aware of another world out there. They
have a growing awareness that there exist other "truths" apart from
their mothers', fathers' or their wider family's point of view. The
teenager opens up to a world of opinions and possibilities, becomes
aware of the world outside their own small world, outside their
family. This awareness causes conflict because the teenager wants to

explore much more of their outside world and imports new views and ways of doing from their newly found truths.

The young adult is now looking for a balance. They try to fill in their own needs whilst at the same time giving consideration to the outside world. They try to balance the happiness found in their surroundings with the happiness they can give. They try to balance the old world, the things learned at home, with the new world, the things being learnt in their wider environment. They try to create a life that is rooted in his past but also developing in their future, away from the roots of the past. They try to find an outside partner who represents all that is new to them, and asks permission from their parents to take this step in life.

Middle age is about finding stability, having established that balance. But at the same time, there is a sense of wanting to change (mid-life crisis). Not only is there, in the physical reality, the stability of knowing who you are, what kind of work you are good at, how you can provide for yourself and for your family, but very often this is also the time for change. The change that is felt, and often pursued, is one that concerns the inner needs. It highlights hidden talents. The stability of income and the routine of life gets balanced with opportunities for another kind of creativity. We are no longer just parents, responsible for the basic needs of children; we are now grandparents who have liberties you don't have as a parent. This stage is about learning to develop creativity as well as living a stable life.

In *old age* we turn more inwardly than we have ever had the opportunity to do before. We consider our own needs first. We also consider our own opinion first. We have a great deal of experience, on which we can base our opinions. We learn when to be kind and when to be hard. We learn when to give and when to stop giving. We learn what we knew when we started this journey; through experience we have learned what is really important in life. We learn about the outside powers that go way beyond the human power, and we learn about the inner power we all have. We learn when to stand up for ourselves and for others, and we learn when to remain quiet and to accept.

These are the seven different stages of development that each human life goes through. These stages are clearly defined and are easily recognisable for their different contents and formats. The lessons are concentrated on different themes in each of these layers. It is as if we enter a different classroom with each stage of our life. Each of these stages has a side that is directed more inwardly and one that is directed more outwardly.

These different stages appear in the same way as all the other seven layers, and they hold the same themes in the same places as all the other seven layers of creation. Have a closer look:

Baby is about knowledge and connection to deep intuitive knowing – number 7.

Toddler is about communication – number 5.

Child is about movement – number 2.

Teenager is about growing awareness – number 6.

Young Adult is about searching for balance in life – number 4.

Middle Age is about finding the right form to contain both – number 1.

Old Age is the power of the self (inner) in relation to the outside power – number 3.

At each level there is a power struggle within the learning process of the yin and yang forces. We experience them in a certain way, each of us in a variety of packages, which will shape our experiences and from that our whole life. If we are shown within a particular development stage that the yang aspect is overriding everything else, then we will grow up having no doubt that there is only *One Way*. In our experience, everybody does it this way, or everybody thinks that way. There is no choice! Apparently. The polarisation of the chakras will ensure that only energies with the "right" orientation will reach the tissues, our inner self. And because we see this so clearly, we assume that the rest of the world does too. *But Everyone Knows…* is a beautiful example of this. The truth is that not everyone knows the world the way you do. You can imagine this to be like polarised sunglasses. On the outside one sees a reflection of the outer image because most rays are reflected back from the glasses, and only rays in a particular direction are allowed through. This minimalizes the strength of the light signal at the back of the glasses, which is the reason for using polarised glasses as sunglasses.

Eventually, over many generations, we as human beings will learn both sides through experience. We will have to move to extreme positions in order to learn that no one position is the truth. It is always a balance of yin and yang, and one that alters with the outside forces all the time. All extreme experiences within human living are necessary for us to learn the lessons we need to learn. All conflicts are necessary. All pain is necessary. Humanity cannot move on without experiencing that no single position is the fixed truth. And we experience this by taking up every single position possible and trying to make it work under all conditions, only to find that it never satisfies. So we should, on an individual basis, not be concerned with making mistakes; they are not mistakes, they are lessons. We need to be allowed to fail our exams in the School of Life in order to progress to the next level! Make your "mistakes", observe the consequences of your behaviour, your opinions, your beliefs, and become aware of the truth. Then you have succeeded!

The formation of the classroom, the way the universe unfolds the various layers of information, is very particular and, by now, familiar to all of you reading this book. Yet, isn't it strange that the way those seven layers present is in a different order from how they are formed? The chakras are numbered from 1 to 7, from bottom to top. They are not numbered by the way they appear or are created. Why would that be?

Obviously, we learn things via our observations of the outside world, and it is in this way that we have noticed the chakras, and everything else of course. So, they were numbered in the order they appeared to us: 1 to 7. It was much later that we got to understand how the universe created the things in it, and then we realised that the creative process of the various layers, and the functioning of these layers, is not the same. What we see here is that the creative process delivers the next layer, which has denser characteristics, out of the previous subtler layer. When the creative process of all the frequency bands within one set of 12 steps and seven major layers has finished, the set is ready to "live" it. *Living* means that experiences are going to be collected and stored. The first experiences are about the characteristics of the first layer and all these will be noted. We first fill in all the spaces with experiences that connect with the aspects of life that relate to the issues of the first layer. Only after we have filled the first layer with the knowledge of our experiences about the issues that are represented within that layer, will life move on to the second layer; this being the second layer of the structure, not the second layer of creation.

This means that humanity has first lived lives that were centred on issues of tribal life, family traditions, safety and security. We then opened up a new perspective in the second layer by living issues that related to becoming individuals, experiencing our own power, including taking responsibility for our sexual power. And it is this stage where the emphasis lies on *movement and mobility* that we are living through in the world at this moment in time. In terms of the developing life, we can see that the development of the human being is at the level of the second chakra. For those who believe that the end is nigh for the human race, this might come as a disappointment, because the road ahead for humans is most definitely still very long and hard. We are a long way away from finalising our growth and development potential. It is only by filing the experiences from bottom to top that any set of seven layers within this universe can stabilise and express all its potential.

Every layer of the six that has been created so far within this universe (six days of creation), is busy filling itself with the experiences representing all the different aspects from "grounding" to "knowledge". So, the various *layers of water and earth, plants, egg-laying animals, mammals and humans* are all busy developing, collecting information through experiences to make the creation "real". It is only when a layer has reached its top layer (number 7) that a new octave, a new set of seven layers, opens up. This means that all existing previous layers of this universe have reached their top development stage before the human being has been created.

In other words, the water and earth layer have developed fully, as have the plants and animals. This means that some species have only lasted for a certain length of time, as they were part of the experience but not of the fully developed product. In that respect, it is a natural process for species to disappear over time. Others will stay because they hold the essence of the experiences that represent that particular development aspect. The crocodile, for example, is one remaining type of dinosaur. The great variety of species will eventually narrow down to a set of essential specimens that are perfectly balanced with their developing environments. It now fits perfectly in the creation as it grows, and those species will remain almost indefinitely.

As a result of the swing of the yin and yang energies on every point of development, that point will eventually settle in its point of balance. Then that particular point will no longer vary in time; it will show a specific expression all the time. This means that the development of that point, the learning of lessons, has reached its final state. It now "knows", with no more doubt, no more trying out. These are all steps in the direction of the developing universe whereby all the created layers will fill up with experiences that have reached their conclusion.

When the human layer has reached its full potential, the universe will be ready to create the seed layer, from which eventually a new universe will be created. As we are still finishing off the second layer of human development, we can be sure that the time for that seed element is still a long way off, at least in human terms. Looking at it from a single human point of view, it is so immense that there is no way we can even have an impression of what that next layer will look like. And, consequently, all that we do notice and see is so infinitesimal that even pretending to understand it illustrates the low level we are still functioning on. The School of Life has seven study years and we are still trying to finish the second year. Yet, sometimes we like to believe that we know it all. That is typical of ignorance; it is expressed loudly and with fervour. We are the small children of the universe right now. Our parents are trying to teach us stuff but we think we already know it all!

Health and Disease

Health occurs as a result of a balance that is achieved around each point of the system. There is, for each point, an optimum pressure. When energies are compressed, frequencies descend into lower octaves. For each step on the way there is a middle point of balance, which "tunes" the frequencies to their most delicate and pure wave movement. In music, this is a pure note; a harmony between all notes and frequencies that makes up the structure, and makes the structure "sound". It is the strongest and will last the longest, within the given remit of the structure (1–2–3–4–5–6–7) because it was built during the creative process (7–5–2–6–4–1–3). This is exactly what we determine as health for our body and mind. Any disturbance away from this balancing point will start the disease process, and more certainly if the disturbance is maintained in the same direction for a prolonged period of time. Our diseases, as described by the medical profession, are only groupings of gross observations into patterns that we have named. Traditionally, people recognised the *Process of Disease* rather than having a name for every possible manifestation. This allowed them to recognise imbalances much sooner and they were much more aware of subtle changes that could bode trouble in the long run.

Anything can cause disturbances to the balance, ranging from external influences such as the weather, food, the people we mix with, the information we receive, to internal influences such as how tired we are, the beliefs we hold, the experiences we have had, or internal clean-up operations that are on the way. In other words, it is constantly on the move! There is no fixed state of health and balance; it needs constant readjusting to the forever-changing environment. However, if our outer- and inner-environments are constantly changing, then that means that the energies that are creating these environments are also constantly on the move, and therefore we are constantly changing too.

What happens to those energies and the way they are moving can be understood by the way radio works. Radio stations broadcast on certain frequencies, but these frequencies are, in fact, a band of frequencies, not one single fixed frequency. As long as your receiver is tuned in somewhere within this frequency band, you can hear the programme. In fact, when you hear the programme in one place you might be receiving it at a slightly different frequency from when you are hearing it in a different place. Each "environment" has its own balancing point, which is the frequency that is best suited to listen to the broadcast, to receive the message in the clearest possible way. That is balance. On either side of this point there is a band of slightly off frequencies that will still give you the same broadcast, but may mean the reception will not be as clear or as stable. Towards the edges of these bands, the reception will definitely be distorted; still the same broadcast reception but hardly understandable now. There will be a lot of interference from other frequencies containing other messages. If you go a little further and you will lose the broadcast all together, eventually you will end up listening to a different station.

When the frequency becomes lower than the one from the balance point, energy is compressed into a higher density. The result of this, in terms of matter, is that the material gets denser, or more compact. It remains the same energetic information when the message remains the same, but it is manifested in a different way. Water vapour in the air, for example, cannot be seen under normal circumstances, and you walk through it without noticing it. However, if you condense the air then water droplets will form, which we experience as rain, and these become so heavy that they fall out of the air that contained them before. Increase the pressure or lower the temperature – to increase the density within the energetic field – and the water becomes ice. These are three stages of the same information displayed at three different levels in balance. They are not the same when the frequency cannot be maintained.

When we look at the seven different tissues as they are created by compressing the human energy field, we have seven different balancing points around which seven tissues are manifested. Remember, the sequence in which these appear towards higher density is from the seed, where juices appear, then blood, followed by muscle, fat, bone, nervous tissue and ending with seeds. Also remember that "vitality" – a measure of how much a tissue is alive – is increased by action (generates internal energy) and decreased by inactivity; hence the saying, *Use It or Lose It.* According to this set-up, the energy of the tissue that cannot hold its balancing point will start to show signs that will look like the next tissue in line.

An example of this is muscle tissue, which is healthiest when it is being stimulated. We know this and have built a whole industry around how we should exercise. When muscle tissue does not get enough stimulation, the frequency that operates the muscular tissue will start to drift away from its balance point towards a denser frequency (a yin shift). In the first instance, this means that the muscle will become harder, less flexible and it will lose operating space as well as power. The muscle will not be able to stretch as far; it will get stiffer. When this drift continues, some muscular tissue will start to show features of the fat tissue, the next layer down; muscle will turn into fat. An increased fat content, especially on the outside underneath the skin, is the direct result of a shift in the energy frequency that normally produces muscle tissue. When this drift of frequency continues, the original muscle tissue becomes bone tissue, and we notice a calcification of the muscle.

On the other hand, when the temperature rises or when the requirement for muscle power diminishes (lowered pressure), the frequency of the muscle tissue increases somewhat and, consequently, the muscle tissue becomes less dense (a yang shift). This results in sloppy, loose muscles without much possibility of putting power into them. When the drift in frequency continues upwards, muscular tissue liquefies; it becomes thick fluid, which we know as pus. This we see in an extreme form in some muscle-destroying diseases.

When the actual physical tissue changes shape and becomes a different tissue, the carrier of this shift, the individual, has serious health problems. However, had we been more aware of earlier signs of this shift away from the balance point for each tissue, we would have been able to rectify the situation much easier. Once again, we see that illness is a gradual process that should be monitored and small changes should be seen to. These are changes within the energetic field first, but have sustained long enough for the changes to have sunk into the actual physical tissues.

It would be best, of course, to notice the changes within the field itself, before these are expressed in physical tissue. For this, we need to be alert to any regular thoughts or emotions we display under given circumstances that result in emotional or mental strain. By altering those thoughts or emotions, we prevent anything from settling into an altered material state. However, it isn't easy to be alert to everything we do all of the time! Our consciousness only has a small capacity. Therefore, if we do become aware of certain reactive patterns within life, and we notice the stress, strain and pressure that these put upon us, it would be wise to stop this automated response programme and to start using consciousness to rewrite the programme. But mostly we will have to content ourselves with becoming aware of physical stresses and strains that are late indications of the shifting frequencies. Take notice of everything your body is telling you! Observe the strain, irritation and pain, and if the system is not able to rectify the problem, to shift the frequency back towards the balance point, we should start to help it. And here lies another problem!

In our heads, we often think we know what is needed, what is right for the system and what is wrong. And most of the time we are wrong! How can we know this? By observing the result of our interference. If people continue to be ill and continue to die after we have treated them in the belief that we can cure them, we have got the treatment wrong. It is that simple. To really know what is required by the system is to really listen to the system. This means: *do as you are told.* If you feel you are very tired, then rest. If you feel you don't want to eat, then fast. If you feel you need to warm up, then warm up. We need to stop thinking, and start *Sensing* the right thing to do. We need to get our thinking, our very small, ignorant mind, out of the way in order to facilitate healing.

In conclusion, we can say that health is a perfect harmony of all the frequencies operating around their balancing points, remaining close to those frequencies, at all times. Disease is any wavering away from those frequencies, up or down, sustained for any length of time, because it will take a bit of permanency before such a change affects the dense layers of the physical matter. *Returning to health is achieved by altering the way of living* that has shifted us out of balance and brought us to disease. The tissue will always follow the energetic input for as long as it has internal power. When it has lost its internal power – its will to "live" – the tissues will collapse onto themselves, which means they very quickly become denser. This in turn means that life itself is being squeezed out of it. Other than that, health depends on our ability to spot changes in our personal energetic field, and most definitely in our physical tissues, and then rectify that part of our behaviour and beliefs that is contributing to the shift in frequencies we are observing. In other words, keep changing with the changes of the outer environment and you will remain healthy.

Health from an Energetic Point of View

It is important to realise that the human being, just like everything else in the universe, is in constant movement; everything is being created at any moment in time and again at the next moment. You very likely experience yourself as the same day after day, but in fact you are not the same even from moment to moment. When you collect some of these fleeting moments then it isn't difficult to imagine you are not the same person you were. Take successive photographs and watch yourself change! This can only be understood by realising that it this is a gradual, never-ceasing process. The changes did not happen quickly in the hours that passed before the next picture was taken; they occurred day by day, moment by moment. This means that the body is continually shaped by the influx of energy coming from the individual's human field. Each one of us has one!

The information, which is in fact a wave or a specific frequency, is being materialised. You could say that the information is written into the tissues. The tissues are formed in a specific way that relates to the energetic information that built those tissues in the first place. The next moment gets written on top of the previous one. When these subsequent moments add different aspects of information to the way the tissues were built, then the tissues become flexible because they are capable of allowing a wide range of possibilities. On the other hand, when a whole collection of moments' information deliver exactly the same message to the tissues, then these become more rigid. They have less flexibility and the "knowledge" they contain will be a limited message.

An example is when a person has always had a message of love expressed by being offered sweets and sugars, then this person will have tissues that long for sugar every time he/she feels a need for love. Another example is when your father was someone you always needed to be afraid of because he terrorised the household, then your tissues, as a child, learnt to cramp up, to harden, in order to absorb the beatings. In general terms, we can comprehend that the manner in which tissues are formed is a direct materialisation of energetic information. The more unilateral this information is, the more compact the tissues will be. Our systems receive information at every moment from all of our experiences, so it becomes obvious that the messages that are repeated most frequently will have a more prominent impact on the physicality of the tissues themselves. These, in turn, will have a greater influence on the formation of the body than messages we receive more sporadically.

Every time we have an experience that is similar to a previous one, a connection will be made within the physical matter to link them up. A new experience hooks up with a previously recorded one that is buried deep within the tissues. This results in a deeper and progressive learning process. We don't need to "get it" the first time, but after a while the larger pile might just set the consciousness alight. The reason they connect is simply because the incoming messages have related frequencies and resonate with one another. This is also the mechanism whereby simple triggers can result in out-of-proportion reactions.

We have learned that visible light divides into seven different colours, each with their own frequency. However, the frequency that produces the red colour is not simply the red frequency, but instead consists of a band of frequencies. It is the same for all the colours. Added to this, the width of the band corresponds to the amount they contribute to the particular colour; hence, we can write it down as a code. When the contribution, listed from largest to smallest, is as follows: red–green–indigo–orange–blue–violet–yellow, then the materialisation becomes visible light (code 1–4–6–2–5–7–3). When the contribution list is green–indigo–red–orange–yellow–blue–violet, then we materialise the colour green (code 4–6–1–2–3–5–7). We will still end up with green even if, for example, indigo has a slightly different hue. In other words, there is a band of frequencies for every colour, even for every band, which won't dramatically change the outcome; it will still be recognised as of the same colour. As long as the code remains the same, the end result in matter will be very similar and recognisable – same tissues. And this is the principle on which our tissues are formed.

Certain information, for example information about communication (carried by the blue frequency, number 5), can present as more compact or less compact, or the Chinese would say, more yin or more yang. This means the tissues can have a more solid structure or a looser structure, less dense. A person with more "dense" communication tissue will be more inclined to listen; the opposite one, the looser communication tissue, more inclined to talk. Both are aspects of the same frequency band, number 5, but they are at the opposite sides of the spectrum. Neither can be called healthy or diseased; it is simply a way of recording information into the tissues. This means that whatever is considered to be "normal" should be related back to how that particular person has been created. "Being in balance", being healthy, will then be measured against the balance of the framework of the individual. Eastern medical systems take the constitution of a person into consideration when determining health and illness for that individual.

Health Is a Personal Balance, Not a Population Average.

During the formation phase of a human individual, when the basic layers of tissues are formed (embryo–foetus–baby), certain information is manifest in yin form – compact and not flexible. Then, later on in life, when this person becomes Stressed, when they have a new information influx that stimulates the function of the tissues, more of a yang nature is formed. A person whose life is built around strict routines and ever-present certainties will become totally lost when changes happen. On the other hand, a person who has recorded early information as yang will become stressed when new information is restrictive and fixed. Both will experience *health* when there is a balance between the deeper, older, tissues (structure) and the more recent ones (function). When the new tunes we want our tissues to play are *In Tune* with the structure of the instrument – our tissues and our body – then we will find peace and calmness in life. The instrument then delivers the purest of melodies, which means that the person is living a life of harmony. One will be more of a leader, while the other needs more guidance and instruction. *Some Chiefs, Some Indians!* This does not make one more important than the other, nor better or worse, just different. Everybody has the opportunity to live a happy life, a balanced life, as long as he/she leads the kind of life the system has been designed for.

Don't forget that life changes all the time, which means that the tunes that are played around us, and that we are asked to play, change all the time too. It isn't difficult to see how our modern world has changed so dramatically within one lifetime. Most of us still remember a completely different-looking world when we were young. Think about no cellular phones, or even telephones, no television, no computers, fewer cars and empty roads, no flying holidays, no "one-does-all" washing machines, no video cameras, no satellite navigation systems, etc. As a result, people are confronted with different "tunes" of the same information, the same frequencies. Their individual capacity to accommodate these determines their individual level of stress and their balance of health.

When a frequency changes along its spectral band it can become tighter of looser, more yin or more yang. This is the way that all frequencies change continually; they vibrate around their balancing point. But when a certain message becomes more and more consistent for a time, or when a certain expression of the frequency dominates, then the inner structure of the person, mentally and physically, needs to alter in order to stay in balance. If the person is incapable or unable to make these changes, then a long-term disease process will start, resulting in early deterioration, and ending in an early demise. The person has to comply and make the changes in order to remain healthy. A "change" of structure, of tissues, will not happen unnoticed. A natural way to notice these changes happening occurs in growing children, where the common childhood diseases are physical signs of such changes taking place, both mentally and physically.

The manifestation is one of physical alterations within the person, within their body, in accordance with the newer information received and upon which action has been taken. Within the growing-up process, different messages are allowed to enter the child's system, to which they then need to adjust their structure. In this light, common childhood diseases should not be suppressed or fought, but instead they should be encouraged and supported because they are clear signs of the child moving into a new phase of life. In order to move up a step in life, the individual has to break down some of the old structure, whereby a lot of waste and debris is produced, which needs to be eliminated. In the first, acute stage of this changeover, it shows itself as a phase of destabilisation, or imbalance, of health. In itself this is a short-lived phase that will automatically and quickly restore itself to a different balance, adjusted to the new environment, and to the new influx of information. Now the individual is ready to face the outer world in a different way. *What We Call Disease Is, in Fact, a New Adjustment to Restore the Balance Between the Inner World and the Outer World.*

When a person loses his/her balance more and more, that is when he/she becomes ill. When left to nature, a reaction will occur in an attempt to restore the balance. But this acute reaction crisis, which we call disease, is something we fail to accept and we want to dismiss. In the natural way, this acute healing crisis, when successful, is followed by a new balance – a period of good health. But if we don't allow this transition to complete itself, bits of debris will be left and the result will be an unfinished structure. When the system is not allowed to clean up properly, the waste products accumulate and the new tissues are not completed. In the end this leaves the structure weakened and carrying a huge amount of waste products. This we know as chronic disease!

In fact, all changes in frequencies that create material substances are a natural phenomenon: it is called *life*. It was happening long before we started to interfere on a large scale and it will continue long after we have been forced to quit interfering. Life itself needs these changes in order to be able to evolve. Life and evolution are inseparable. *Trying to Keep Everything the Same* is a recipe for disaster. We *Need* to change, and the body needs the freedom to go along with these changes. Changing frequencies only slightly – recognisable information but presented differently – over a long period of time should enable the tissues to adjust their physical form and/or their function to maintain balance. Balance means health, so whilst adjustments take place inside the body we will notice them. We call these signs the symptoms of illnesses. In the Western world we have done away with the idea that nature restores everything, that there is such a thing as an internal natural healing power. So as soon as the first signs of a change appear, we want to *Do Something About It*. Doctors tell me, as if they know for sure, what my blood pressure should be at this particular moment of my life, under my specific circumstances, or what my blood sugar level is, or my lung capacity. They take an average of a group, decide on the conventional way to approach the measurement, and then pretend that everybody should be like that, and at all times!

There is no room for change in the Western world! Frequencies, however, vary all the time and move from one side of the scale (yin) to the other side (yang). This confronts us with a complete set of information encapsulated within the frequency. It is because of the contact with the full range of possibilities that we realise that there is a choice, that there are different possibilities we can choose from to make the balance in life different from what it has been. The combined mass of possibilities of the seven most important frequencies shows us the variety of human expressions that are potential choices. No two are identical. This means that whatever someone else has experienced will be different for us, even if the circumstances and the events appear the same. No two rain showers are the same. The way we are influenced by the circumstances is different from one person to the next, but also from one moment to the next. You handle the same situation differently when you are 19 than when you are 39. In essence, it is the same difference you may notice between last night and this morning! Maybe it depends on who you spend the night with! *Health, Balance in Life, Is Invariably Connected to the Moment.* There is no routine, no therapy, that can guarantee your health into the unforeseeable future, because the "routine" cannot know what information you need to adjust to in that future. Now that we know this, how can we redefine health and disease?

The tissues of the foetus are formed as a result of the influences and the information that is compressed from the energy field into matter. This process "fixes" information into matter. You could say that the composition of the tune has been recorded in the studio; it sticks to the vinyl (another something that has changed). Then it is the baby's turn to collect information and fixate it. In respect of the information that the baby collects, and that confirms the foetal information, the tissues will be strengthened. Information that does not fully comply, or notes that are received differently, are integrated into an expanding frequency band that is noticed by the tissues. Totally different information is now causing stress for the baby. Maybe the world looks very different in direct experiences than it did according to the mother's information. It could be as bad as it being almost impossible for the baby to function within the world, to survive the onslaught of "unsuitable" information waves, dramatically disturbing the tuning of the tissues. This leads to disease, which is a malfunctioning between the structure of the system and the way it is being played, and the way it is being asked to perform. This kind of environment is not feeding the child, it is not supporting the child, and it is not stimulating growth and expansion of the structure of the child.

   Illness is a manifestation of a long-term exposure to an environmental tune that differs vastly from the one stored within our
tissues. For a short time, on the surface, we can easily adjust, a bit
like we do every year when the summer season turns to winter.
But things that cause a deep-rooted discomfort, things that go
way back in time, begin to alter the function of the system; first
almost silently, later more persistently. We take very little notice of
it, because we are too busy and we only want it to go away. We continue on without allowing our systems to make some alterations
to our structure. We don't want to feel "bad", so we ignore the first
calls. And we ignore it for as long as we can until we are forced to
listen because systems start to collapse, or because a system stops
us from carrying on. We are being invited to change our life. And if
we do change, it is mostly because we have "give in" to the disease,
not because we have understood that the system required a change.
Doctors may tell us: "You will have to learn to live with it!" Still
no need to change. The end result is chronic illness, which is long
drawn out with no prospect of it ever getting better.
   The other path to illness is exposure over a very long period
of time to a *Never Altering Tune*, which becomes out of tune with
the rest of the changing frequencies. This puts the tissues under a
great deal of strain in order to try to hold on to the old structure.
"Holding on" means that tissues harden up, condense, or become
so compact that there is hardly any flow of energy left through
the tissue. This manifests in serious, chronic and life-threatening
diseases.
   The healthy option is to listen to the body that tells you that life
isn't easy and that you better not continue this way, that things need
to change. If you listen well, you decide to live differently. How?
What is it that needs to change? Before you can figure out the
answer, you need to stop your old habits or old ways of doing things.
One immediate way to start this is with the way you use food. You
need to stop clinging on to old frequencies. So, first you must stop
what you are doing. Sometimes it is obvious which direction these

changes need to go in, but at other times we can only question everything we do in order to find the right new direction. In that case, stop as much as you can, and – very importantly – make sure some essentials of your current life are included. For example, you could temporarily stop working, or stop your hobbies, or take time out from friends, from your partner, from your children.

From the moment you create space, from the moment you disconnect the impulses of the outer world and your automated reactive patterns, you may feel a sense of relief. Whether you do or you don't, however, the main point is that *The Pattern That Is Making You Ill Has to Change.* You must realise that it is the way you have lived up until now that has brought you to this imbalance and state of illness. Creating health from this point onwards means, in the first place, to stop living your life based on old thought and feeling patterns. Once peace returns to the inner world, information about what exactly has been so disturbing and disrupting will float to the surface. Amongst this you will find most of the answers to your earlier questions. These answers emerge from the depth of your being; they are not the result of long and hard thinking. When you act upon these insights, and you make the appropriate alterations to your life, the tissues will move closer to the balance that is being asked for by the structure of the system. *The Road Back to Health Has Natural Signalisation and Is Paved with Intuition.* It boils down to the fact that we need to embrace a variety of frequencies, of tunes, of music, in order for us to realise which frequencies the inner world requires for balance, so that we can arrange for those frequencies to be present in our outer world. A tuba that wants to sound like a violin will break itself trying!

The music that the inner structure likes has deep roots in the system itself. Our structure resonates with the music we ourselves produce. However, do not forget that your inner tunes also alter over time, with experiences; hence, *Your Health Is Determined by Your Potential and Your Willingness to Remain Tuned in to Your Own World.*

Make music together, you and the world.

Play in harmony.

Know that your tune will die away long before the world's does.

Chapter 9
The Cartesian Coordinates

A Cartesian coordinate system is a way of identifying any point in a given space. It is rectangular in form, whereby the distance between two lines is constant. For every dimension there is an axis and the axes meet perpendicularly. The set of points within this system, which are identified by their coordinates in relation to the axes, together make up the Cartesian plane.

It is the system that is used the most in mathematics as well as in physics, because in this system geometrical matter can be placed most easily. It was the French mathematician and philosopher René Descartes who first came up with the concept. His Latin name was Cartesius.

In 1637, Descartes developed the idea in the following publications:

- *Discours de la Méthode*: In part two he introduced the idea to identify the position of a point within a plane by using two perpendicular crossing axes as reference.
- *La Géométrie*: Here he developed the idea further.

Choosing a Cartesian coordinate system for a one-dimensional space – a straight line – means choosing a point o on that line, which represents the origin, and choosing a unit of length and an orientation of the line. The latter means choosing which of the half-lines is *Positive* and which side is *Negative*. We then say that the line is orientated from negative to positive. Any point, P, on the line can be specified by the distance to o, coupled to a + or - sign, depending on whether it is placed on the positive or negative side of the line.

A line with a chosen Cartesian system is called a *number line*. Every real number, be it a whole number, a rational or irrational number, has a unique location on that line. Looked at the other way around, every number on the line can be interpreted as a number in a coordinated continuum of real numbers.

Two Dimensions

A Cartesian coordinate system in two dimensions is defined by two axes meeting perpendicularly. The points within this system form a plane; the xy-plane. In drawing these axes, it is conventional to draw them horizontally and vertically. The horizontal axis is called the x-axis and the vertical is called the Y-axis. The crossover point is the origin, marked by o. Each axis will be divided in points equal distances apart, which gives us a measuring unit. Every specific point of this plane can now be identified by two coordinates (X,y), representing the direct distances from the particular point to both axes.

Here are some examples. The points $(3,4)$, $(-3,2)$ and $(2,-2)$ are represented in the following drawing. The point $(0,0)$ is the origin.

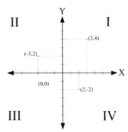

The arrows on the axes indicate that they go on indefinitely in that direction. The two axes create four quarters, indicated by Roman figures I, II, III and IV. The quarters are marked anti-clockwise starting with the upper-right quarter. This allows us to identify every single point in a horizontal plane in relation to the chosen origin.

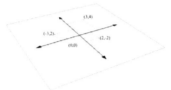

The following table shows the values of the x-axis and Y-axis of the quarters.

QUARTER	x value	Y value
I	> 0	> 0
II	< 0	> 0
III	< 0	< 0
IV	> 0	< 0

Three Dimensions

In the early 19th century the system was expanded to three dimensions. For this purpose, a third axis was introduced; the Z-axis. By doing this we are capable of moving from a plane (two dimensions) to a space (three dimensions). A point in a three-dimensional space can be found by the coordinates (x, y, z).

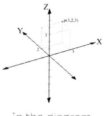

In the diagram above, point p(3,2,3) is depicted.

The Cartesian system divides the three-dimensional space into eight parts.

The first part is located *"Above"* quarter I:
 The values of the x-axis, y-axis and z-axis for point P (any point within this first part) are x > 0, y > 0 and z > 0. The three coordinates in this part all have positive values, which can be written down as (+x, +y, +z). Let's call this part **Ia**.

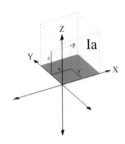

The second part is located *"Above"* quarter II:
 The values of the x-axis, y-axis and z-axis for point P in this second part are x < 0, y > 0 and z > 0. We can identify P by (-x, +y, +z). Let's call this part **IIa**.

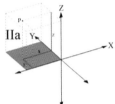

The third part is located *"Above"* quarter III:
 The values of the x-axis, y-axis and z-axis for point P in this third part are x < 0, y < 0 and z > 0. We can identify P by (-x, -y, +z). Let's call this part **IIIa**.

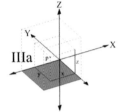

The fourth part is located *"Above"* quarter IV:
 The values of the x-axis, y-axis and z-axis for point P in this fourth part are x > 0, y < 0 and z > 0. We can identify P by (+x, -y, +z). Let's call this part **IVa**.

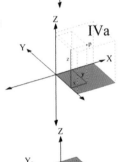

The fifth part is located *"Below"* quarter I:
 The values of the x-axis, y-axis and z-axis for point P in this fifth part are x > 0, y > 0 and z < 0. We can identify P by (+x, +y, -z). Let's call this part **Ib**.

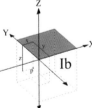

The sixth part is located *"Below"* quarter II:

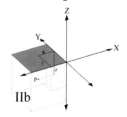

The values of the x-axis, y-axis and z-axis for point P in this sixth part are x < o, y > o and z < o. We can identify P by (-x, +y, -z). Let's call this part **IIb**.

IIb

The seventh part is located *"Below"* quarter III:

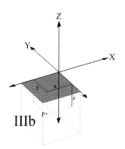

The values of the x-axis, y-axis and z-axis for point P in this seventh part are x < o, y < o and z < o. We can identify P by (-x, -y, -z). Let's call this part **IIIb**.

IIIb

The eighth part is located *"Below"* quarter IV:

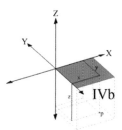

The values of the x-axis, y-axis and z-axis for point P in this eighth part are x > o, y < o and z < o. We can identify P by (+x, -y, -z). Let's call this part **IVb**.

IVb

The Human Body in a Cartesian System

We learned from our discussion about the colours of the visible light spectrum that we need to take notice of the various codes. We learned that indigo can be created in two different ways and that orange has four possibilities. Put those together and we have eight different schemes. We have been wondering for years why it actually is that way.

We have related the various codes to specific expressions in subsequent organisms and we said that 6A shows up in the bottom part of the organism (roots) and 6B in the top part (plant). Equally, we find 2A to be the front (6–5) left (4–7); code 2B the front (6–5) right (7–4); code 2C the back (5–6) left (4–7); and code 2D the back (5–6) right (7–4).

Putting these together, we are able to divide the human body into eight parts too. This could be done as follows:

- the YZ-plane divides the body into a left and right side;
- the XZ-plane divides the body into a front and back side; and
- the XY-plane divides the body into an upper and lower side.

Do you remember from matching the light spectrum to the physical body that we discovered that the middle of the body was coded 3.3.3?

From the composition of the colours within the spectrum of visible light, we select code 6B and code 2D, which is part of the representation of the human being. A stands for the spectrum of light, B for the human field and C for the individual field. This gives the following representations:

- Part 1 and part 2 – front right:

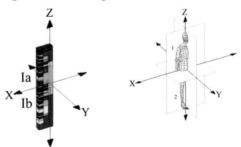

This area covers "**Ia**" and "**Ib**" in the Cartesian coordinates. Each point p in "**Ia**" has the coordinates (+x, +y, +z). Each point p in "**Ib**" has the coordinates (+x, +y, -z).

All the points on the right front of the upper body we should be able to identify in the same way: +x, +y, +z. And those on the right front of the bottom part of the body have the coordinates +x, +y, -z.

- Part 3 and part 4 – front left:

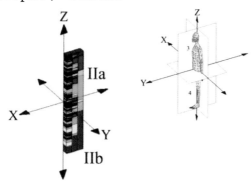

This area is compatible to area "**IIa**" and area "**IIb**" of the Cartesian coordinates. Each point in "**IIb**" has a coordinate (-x, +y, -z). Each point in "**IIa**" has a coordinate (-x, +y, +z).

All the points on the left front of the upper body we should be able to pinpoint in the same way: -x, +y, +z. And those on the left front of the bottom part of the body have the coordinates -x, +y, -z.

- Part 5 and part 6 – back left:

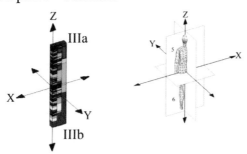

This area is compatible to area "**IIIa**" and area "**IIIb**" of the Cartesian coordinates. Each point in "**IIIa**" has a coordinate (-x, -y, -z). Each point in "**IIIb**" has a coordinate (-x, -y, -z).

All the points on the left back of the upper body we should be able to pinpoint in the same way: -x, -y, +z. And those on the left back of the bottom part of the body have the coordinates -x, -y, -z.

• Part 7 and part 8 – back right:

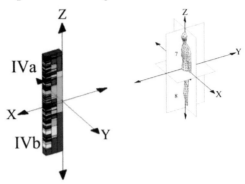

This area is compatible to area "**IVa**" and area "**IVb**" of the Cartesian coordinates. Each point in "**IVa**" has a coordinate (+x, -y, -z). Each point in "**IVb**" has a coordinate (+x, -y, -z).

All the points on the right back of the upper body we should be able to pinpoint in the same way: +x, -y, +z. And those on the right front of the bottom part of the body have the coordinates +x, -y, -z.

Together with the drawings of the human body in the Cartesian coordinate system, we have made a comparible drawing of the division of the visible light spectrum.

We know that the energetic human field shows itself as a physical person and so the big question now is: "Can we connect the points of the human body to its energetic human field, and are those connections of the (x, y, z) coordinate order?" The answer to this question you will find in Part II of *Why Me?*

Chapter 10
The Human Evolutionary System

In the book *Make Your Heaven on Earth*, subtitled "Discover your own wisdom level", published by Akasha (only in Dutch), William Gijsen tells how he lost faith in the Catholic doctrine and became interested in the supernatural as a result of a vision he had of his deceased mother. The change happened after he experienced several medium sessions and he was contacted by a spirit named Anthon. In clear but fascinating language this higher soul explained how the spirit world was structured, where in the body the soul lived and how the human individual helped to shape his/her own life. Anthon talked about evolution, reincarnation and the accumulation of wisdom. Gradually he taught the author, and through him the reader, that every human being has the potential to be happy and successful, if only he learns to use the intuitive good feeling of the soul. In that respect, the book evolved to become a manual for the soul.

In the second part, Anthon talked about the true purpose of the soul. He explained the evolution through seven spheres. For the first time it was mentioned that these spheres were also usable on earth, and how a soul could learn within each sphere and on each level of each sphere. He showed the reader that heaven is "within reach", and how to get there by listening to the soul informing us continually about the future. This book showed the reader at what level of wisdom he/she had arrived (which sphere and which level) and what that wisdom specifically meant to the individual. This knowledge teaches that every earth soul, through intensely occupying itself with the realisation of the sphere task, can elevate the soul to a higher level of wisdom, even here on earth.

Our evolution as humans evolves through various spheres. A sphere is a specific level of wisdom that holds all souls with the same insight in how to create harmony. Altogether there are seven spheres. The bottom three spheres are occupied by humans that maintain a body. In the upper four spheres, souls gather that no longer incarnate on earth; they remain in the spirit world.

Every sphere is divided into seven levels, which totals 49 levels of wisdom. Every level is about learning a particular lesson. Because there are three spheres that relate to earth life, it follows that those humans need to pass through 21 levels of wisdom. Every sphere teaches one big theme of life, and every level then teaches different aspects of that theme.

In the first years of earthly life, a child is still aware of why it came here and what experiences it needs. But because of adaptation to parents, educators and society in general, almost everybody loses that insight. Forever more they turn away from the feelings inside their own soul, which means that by the start of puberty contact with the soul has been lost. In the first years of life intuition is still sharp and acute. Then our logics develops more and more and we rely on that, so we lose our real intuitive power.

When we die the soul returns to the spirit world. There we reside in the same sphere as we did on earth but without a body. In between lives we linger in the same sphere in the spirit world.

Summary: Create Your Heaven on Earth

- There are 7 cosmic grades.
- Every cosmic grade divides into 7 cosmic spheres.
- Every cosmic sphere divides into 7 cosmic levels.
- Every cosmic level divides into 7 cosmic phases.
- Every cosmic phase divides into 3 stages.

In Gijsen's book, the system of wisdom is compared to our own education system, whereby the sphere corresponds to the study direction, the level to the year you are in, and the phase to the specific subject. The stages compare to the chapters. The sphere and the levels indicate the purpose of that particular life. When you live according to your sphere and level, you will encounter the necessary experiences and accidents in order to bring the lessons to a fruitful close, giving you heaven on earth. In our educational system, you are taught to count to ten in the first year at school, how to read and write and so on. If you try to teach these children complex trigonometry and scientific writings about gothic art, you won't get very far. A child at the first level can be taught the lessons that are incorporated within the first year of school.

The stages tell you how to use the skills (the "how").

The phases tell you which skills you need to improve (the "what").

Stages and phases together is the "why".

- Sphere 1 is about surviving (taking cues from someone else).
- Sphere 2 is about adjusting to disappointments (live for yourself).
- Sphere 3 is about superseding (live for yourself and others).

- Level 1 is about manifestation. Key word: perseverance.
- Level 2 is about possession. Key word: enrichment.
- Level 3 is about feeling. Key words: warm sensation.
- Level 4 is about loving. Key word: love.

- Level 5 is about order. Key word: perfectionism.
- Level 6 is about insight. Key word: intuition.
- Level 7 is about devotion. Key word: belief.

- Phase 1 is persevere versus dawdle.
- Phase 2 is to enrich versus to impoverish.
- Phase 3 is warm sensation versus cold sensation.
- Phase 4 is open-heartedness versus heartlessness.
- Phase 5 is order versus chaos.
- Phase 6 is intuitive thinking versus cognitive thinking.
- Phase 7 is religiousness versus wickedness.

- Stage 1 is gathering versus dissipating.
- Stage 2 is clarifying versus obscuring.
- Stage 3 is elevating versus downgrading.

Sphere 1

In this first sphere, the soul is learning to use all of its creativity in order to survive. It learns to decide for itself. It has to learn to survive amongst all the other souls with a free will and with self-reflection. Here the soul often feels the victim of its environment or its situation. Or, it may be the victim of its expectations. The soul assumes that the environment is going to be loving and caring in the same way it wants to be, but it doesn't know that the environment is not yet ready for that. These souls are searching for security; they want to know who they can rely on and who they can't. Hence, the souls divide the world into two camps: for and against. Here, you learn to conquer your place within the group and to maintain it. The central theme is caring about others and the outer world is important.

Sphere 2

In the second sphere, the soul learns to deal with adversity. It learns to make choices, more so than in the first sphere. What are the choices to make and how does it feel about them? How can it put forward conditions surrounding its choices? Here, desires are to be refined. There is a search for motivation and connections. A soul tries to gain more insight into its own behaviour. It is important to learn to deal with your free choice and to strive for independence. The central theme is self-care. The inner world is important.

Sphere 3

In the third sphere, the soul learns to supersede the polarity. Having to choose out of opposites in sphere 2 changes here into learning to find a new creative way of solutions. The soul starts to realise that life is not an or–or but an and–and situation. The soul learns that choosing is not beneficial if it is out of self-gain, but that there can only be real gain if all sides improve. Here, we realise that we should not be following a judgement; not our own or anyone else's either. The central theme is caring for yourself and for others. The coming together of the inner and outer world is important here.

After I read the book *Make Your Heaven on Earth*, followed by some workshops with William, he suggested I examine the pyramid form. The result of this was that I moved up a step and, funnily enough, I did notice it. I went from 2.1 to 2.2. To be more precise, I moved from sphere 2–level 1–phase 7 to sphere 2–level 2–phase 1. In 2.1 you learn to choose, which is not always easy because you are afraid of making the wrong choices. Therefore, often you are stuck in doubt and hopelessness. However, your own choice at that moment is always the right one. You need to learn to be independent and to make your own decisions.

Every level also has seven phases. In level 2.2 you learn that the decisions you made in the previous level need to be executed and fully supported. You now own your choices. Take note of your desires and try to follow them. These can relate to a variety of areas, such as material things, emotions, sexual or spiritual. Because you own these desires you need to try to fulfil them. Then you start to weigh them up and ask: "Why do I want this? What is my purpose? What added value will my life have once these have been fulfilled?" You learn to put them into perspective. You ask what are the main themes and what are the secondary themes? In the previous level we became mad with doubt, whilst here all efforts are geared up to reach satisfaction. It is the result of narrowing it down to one choice and to go for that choice, because it is worth it. This is how you increase your self-worth. Here, you become prosperous because you follow your inner values; the value you give yourself. This filters everything out for you as you decide what is important to you.

You can compare this system of consciousness levels to our schooling system, or the levels you find in computer games. In preschool there are various grades, then there is junior school with various classes, followed by college, university and so it continues into the working sphere too. Each school has children of a certain age group and development stage, and each class has its own teaching programme. You can only move up a class when you have learned the lessons from the one you are in. In computer gaming, you need to complete one level before you can move up to the next level. You need to have solved all of the problems that have been placed in front of you before you can tackle the next set of problems. It is the same with the consciousness levels. Here, you also need to have learned all of the lessons, and to have passed all of the exams of one phase, before you can move up one level.

I believe you can find this system of consciousness levels or wisdom levels in the division of the visible light spectrum. The codes we have so far discovered can be compared to the various spheres, levels and phases. I also believe that a malfunction of a certain part of our body is telling us about the lesson we need to learn, although people don't realise this. For instance, when I have a headache I don't, like some people do, take a painkiller. Instead I'll ask myself: "Why do I have a headache? What am I filling my head with?" Similarly, when my stomach hurts I think: "What am

I not digesting?" With a sore throat I ask: "What am I not saying?" These are only small lessons, but ignoring them makes them pile up and they can then turn into almost insurmountable obstacles.

In the meantime, we are already aware of the fact that our body is a manifestation of energy. This energy we approached previously as an energy pattern $(Ep = Mp.c^2)$, in which we placed the various chakras according to the Golden Ratio. Combining this with the division of the visible spectrum, we showed where the seven chakras presented themselves within the physical manifestation and also how each energetic level had its place.

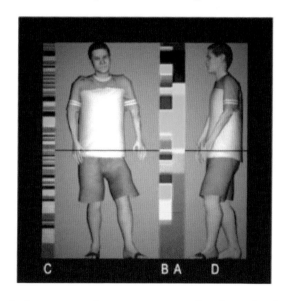

Now, let's compare the consciousness levels to the division of the light spectrum.

We could argue that the first human phase looks like this: sphere 1–level 1–phase 1. Or, 1.1.1 for short. Let's agree that the first figure (1.1.1) stands for the type of education: prep, junior, college, high school, university. In the picture above, this is column c. The second figure (1.1.1) is the class we are attending. In the picture, this is column b. The third figure (1.1.1) is the course we are studying. This we find in the picture as column a.

Now we can create a code system starting from 1.1.1, which is then followed by 1.1.2, 1.1.3, 1.1.4, 1.1.5, 1.1.6 and 1.1.7. The changing figures represent the various courses within the same grade.

These lessons can be found within the structure of the body.

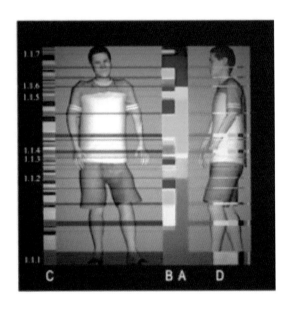

In fact, we don't need the figures. We can use colours to represent this.

When we are finished with 1.1, we step it up another level, from the first year in prep school to the second year. We move up from 1.1 to 1.2. In this class, 1.2, we have another seven subjects to learn, coded as 1.2.1, 1.2.2, 1.2.3, 1.2.4, 1.2.5, 1.2.6 and 1.2.7.

Then follows classes 1.3, 1.4, 1.5, 1.6 and 1.7. The second sphere is made up from 2.1 through to 2.7. And the same goes for spheres 3 through 7.

Let's give an example and find the respective points within the spectrum. Take the following codes:

These codes can be found within the division of the light spectrum.

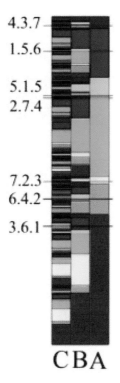

Let's do another one with different codes:

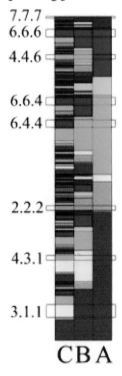

And, again, their corresponding places within the spectrum:

Now, some of you may be shaking your head asking, "Why would you do this?" or, "Where are you going with this?"

The way we have divided the spectrum and projected it onto the body throws horizontal lines across the body. Let's now link this knowledge to the system of consciousness levels. Grabbing some painkillers to relieve the symptom can be replaced by asking the question *"why?"*: "Why do I have a headache?" Or, we can rephrase this question as, "What am I stuffing into my head?" We can make the metaphorical phrases become real.

And we can take it a step further. What about life-threatening diseases? Is it possible that the body is telling us what consciousness level we are stuck at? Learning a lesson may also mean that we need to "unlearn" things we have believed in for a very long time. And that, too, can be shown to you by your body.

Karma

Karma is a term that comes from Hinduism and Buddhism, and literally means "action" or "seed". Both explain karma in terms of the physical and mental actions of the individual, which have consequences in this life and in the next by way of reincarnation. For everyday use we can say that all we do, think and say will find its way back to us. Karma deals with the action, and the consequences of the action are called *"Vipāka"*.

Karma and the consequences are seen as a natural phenomenon, the law of "action and reaction". Here there is no need for a God who punishes or rewards. A good deed naturally leads to good consequences, and bad deeds lead to bad consequences, all in a natural manner. Karma itself does not judge, it has no "good" or "bad". It is simply that deeds of a certain kind will be "repaid" by deeds of a similar kind. This play of action and natural reaction is reflected at every level of the consciousness evolution too.

All religions that include reincarnation stretch this action–reaction play out over several lives, if need be. You may be confronted with the consequences of your behaviour in this life, or sometime later in one of your following lives. Similarly, what happens to us now may be the result of something we have done in this life or in a previous one. Everything we do now will have an influence on what will come to us later. By way of our deeds we are responsible for our own lives, our own pain and our own happiness.

Karma in Buddhism

The Buddha taught the importance of deeds and thoughts. He taught that there are several different types of deeds. He emphasised the two main groups of good and bad deeds by pointing to the consequences, good or bad. The other important distinction is *How* we perform these deeds. There are basically three ways to act: through the body (physical), through speech (verbal), and through thoughts (mental). The mental aspects weigh the heaviest, they have the most impact.

In more detail, Buddha gave a list of 10 types of actions that have both a good and a bad variant. The three main reasons behind bad actions are greed, hatred and ignorance, which means that the opposites encourage good deeds. The opposites are unselfishness, love and insight.

Buddha never enforced his teachings onto his pupils. He left it up to them to find out for themselves. He did, however, teach them the four truths. In these, he showed that someone's deeds are more important than their beliefs. In order to benefit from good deeds, you do not need to believe in reincarnation.

Karma in the West

In the West, karma was introduced in theosophy. This is the "teachings of consequences" that human souls have to learn. With the introduction of yoga and meditation techniques, the knowledge about karma entered society at a deeper level. Be good and others will be good to you; behave badly and there will be a price to pay.

Tornado

It is maybe by being confronted with our karma that we will wake up to the lessons we need to learn. Karma is not a punishment; it is a means for growth, insight, consciousness and love. Sometimes only suffering will bring us to our senses. Karma is a principle of *Self-Development* in which new opportunities for growth and learning are being offered ceaselessly. You choose whether you want to take the opportunity or not. Here is your chance to sow now and harvest later.

Following the laws of action–reaction, a long chain of continual consequences leads to the idea that one thing leads to another and hence there is a purpose to all of this. We have come into this life to learn certain lessons, to follow a certain path. Negative karma can then be interpreted as a sin against oneself and against the outer world. Positive karma is being good to oneself and to the outer world. These effects are taken up by the energetic system. But this also means that within each of the colour frequencies, these interferences will be noticed as, through our deeds, we either underpin the natural balance, the resonance of the system, or undermine it. Then the question is: "When will we be confronted with this karma?"

The answer is truly simple. Karma will be seen at every level it relates to. You could simply say that red karma will show up in the red area, orange karma in the orange area, and so on. It is everywhere, but that does not mean that you are aware of all of it. You will only be confronted with the karma at the level you are dealing with right now, in other words, the lesson you need to learn in the class you are currently in. We need to learn to let go of a number of things in the first place, before we can replace it by different behaviour, or even opposite behaviour. A small example is when I used to mow the lawn in a circular, anticlockwise motion, which left me at the end still collecting part of the cut grass. Now I mow the grass in a clockwise motion and there is no need to gather left-behind cut grass anymore. This is a consequence of the rotation movement of the blades, which are now able to direct all cuttings into the collector. It is the same task, but the consequences of my action leaves me with a different result because I have changed the way I perform the action.

Some people seem to move fast through the lessons whilst others struggle a lot. But you can rest assured that we will all be confronted with it at some point. Sometimes these confrontations will be hard, and if it keeps happening, it will get harder and harder each time, until you realise you don't really have a choice but to act. You really need to make the change. It is a bit like a tornado, which is always the result of warm and cold air having a brutal collision. The warm air expands and the cold air shrinks, the clouds mix together, then the cold air drops down and twists and warms up, which drives it in a twirling motion back upwards. This kind of thing happens in life too.

In a tornado there are three places you can find yourself. These are outside the tornado (A), in the stormy part of the tornado (B), and in the centre of the tornado (C).

A: It is safe on the outside of the tornado. At the moment there doesn't seem to be any danger. A lot of people are situated outside of the tornado. They carry on and they go about their business as usual. They rely on their knowledge and go with the flow.

B: In the tornado itself it is heavy going. You have no more choice; you've got to get through it. It is very difficult to get out of it but you will have to try to survive it.

C: In the centre of the tornado is the most still place you can imagine. Here everything is peaceful.

The people on the outside (A) are in what can be described as "purgatory". They view the situation from an outsider's point of view; they are not directly involved in the storm itself.

The least safe place to be is right within the tornado (B). This is "hell". There is enormous tension here and the suffering is tremendous. The only way out is towards the centre (c) where there is peace and quietness. This is what we call "heaven". Getting there, moving from the storm to the centre, is a kind of *rebirth*. However, the euphoria of getting there is short-lived as we figure out that heaven turns out to be a new purgatory, the one from the next octave. After some time, we will be sucked towards and gradually into the next tornado. This is called "evolution"!

In Dante's *Comedy Dell Arte* we find a description of these three steps. He takes us on his fictitious journey into the afterlife, moving from hell into the mountain of recovery and then into heaven. Originally, Dante called the work *My Comedy*. It was Boccaccio who renamed it in 1360. In the Middle Ages, any story that ends well was a comedy, as opposed to a tragedy, which did not end so well. It had, in itself, nothing to do with humour.

Even the life of Jesus Christ reflects those three stages: teaching (A), suffering (B) and resurrection (C).

In John 3.3 we read: *"Jesus Answered and Said to Him, 'Most Assuredly, I Say to You, Unless One Is Born Again, He Cannot See the Kingdom of God'."*

It is all about surviving the storm. But what, truly, is surviving?

Surviving means:
- still being alive after the event;
- living longer than someone else;
- coming through;
- staying alive in spite of life-threatening circumstances;
- maintaining;
- getting through difficulties; and
- being stronger.

So, surviving has a literal meaning: not dying, and a more metaphorical one: maintaining, getting through difficult times. Now we know that diseases are, in fact, the reaction or the result of something. Not resolving the cause of the reaction is not going to change anything in the long run. However, once you realise what the cause is, the solution becomes crystal clear. An alcoholic, for example, needs to realise what the problem is that is causing him/her to drink, otherwise they will remain an alcoholic. Only when a person knows and accepts the problem can they work at resolving it. And, again, this journey, no matter how easy or difficult it might be, begins with the question *"why?"*.

Even though we have a picture about how human evolution works, we still need to find a way to relate it to our everyday lives here on earth. Let's take the hype surrounding the December 21, 2012 date as an example. We were told it was going to be the end of the world, the "end of all times". To me, it simply meant that humanity had entered a new phase. We moved from the era of Pisces to the era of Aquarius. We have the human code, 7–5–2–6–4–1–3, and we know that every human lifecycle also follows the same pathway, then it seems logical to assume that the whole of humanity also follows the same route. In that coding system it becomes obvious that humanity is now changing from 5 to 2. Frequency 5 is about communication and 2 is about movement. Communication truly started with the advent of the printed word and now the times of telephones, radio, television, and internet, until it has reached the stage whereby we no longer need cables or antennae. We use satellites and we are able to reach almost anywhere in the world at any given moment in time.

The evolution the telephone has gone through serves as a beautiful example. First, we needed a manned panel where somebody had to make the connection for you. Second, it changed into automated boxes, but where you still needed to pre-book a long-distance call. Next, it moved to a system that connected all telephones in the whole world, reachable from your own handset, and finally to where we now don't even need any cable connections anymore and we can be reached from anywhere in the world because we carry the phone in our pocket. At the same time, we have changed all information from printed form to digital, from matter to energy, which again can be reached from almost any given point on earth at any given time. Everything moves from simple structure and complex movement to complex structure and simple movement. In physics' terms, we can state that we initially have little matter energy and lots of kinetic energy, which gradually evolves to become lots of matter energy with little kinetic energy.

People might say that nothing has changed since December 21, 2012, but is this really the case? Remember the basic formula $Ep = Mpc^2$, which states that when the frequency changes then the mass changes too. It is only a matter of time before we start noticing it. It is that which we see on a grand scale in the world today. In the last one-hundred years our technology has changed at a dizzying pace, whilst the human being has almost stayed put. What is happening is that the people who are positioned outside the tornado are now going to be confronted with it, because the environment we live in has changed. We are being sucked into it. It is going to be a time for survival. Not because we want to, but because we have to. We will be facing an economic downturn, which will go hand in hand with a spade of "new" illnesses. Keep an eye out for these simple first signs. Do you see them heading your way?

And what then is the answer? There is only one possible solution: we need to redraw the basic structure of life. Looking around, I notice that "the system" has put us all in the same class, which reminds me of the first sphere; the whole is important, not the individual. In this class we find babies, toddlers, adolescents, adults and the elderly, all attending the same lessons. But this always leads to tension because people require a custom-made programme. It doesn't make a lot of sense to send a seven-year-old child to university to study. Yes, we need a challenge in order to learn, but the challenge should be within our personal achievement sphere. Compare it to a computer game where the virtual reality is divided into levels that form a challenge to each player. Most levels are created by level editors. These are computer programmes that make the various levels manifest. A level very often consists of loose

parts, such as music, 3D objects, textures, and so on. It's the level
editor that brings them all together. Other levels can be created by
a level generator that composes the game randomly. This results in
a continually changing game whereby it is different each time you
play it. A player can only move up a level when he has successfully
completed the previous one.

Now compare that system to the ancient myths. The hero
always had to overcome a massive obstacle, solve a riddle, escape
from a labyrinth or kill a dragon in order to reach the treasure; to
find wisdom. It is like moving through the storm, through the high
winds, to reach the peaceful centre. It is here where the "treasure"
lies; the wisdom and knowledge that will allow you to move to
the next level. It is a rebirthing process, ready for the next task.

The whole universe is about energies, some that manifest in
matter and others that don't. Maybe you realise that God is not
an old bearded man looking down on creation and ruling it with
power. For some time now that story has not satisfied, which means
that God must be something else.

I believe that God is an immense energy field, which holds all
manifestation of energy in it (Chapter 2). This leads me to conclude that God is everything we observe as well as everything we
do not observe. All we see of God is everything within the visible
spectrum. The rest to us is philosophy. Humanity is part of this
immense energy field. Because the universal adagio *So Large so
Small*, you may imagine a human body to be the representation of
the god energy and the stomach could then be the human being.
When there is something wrong with the stomach – it may hurt
or malfunction – this will have an impact on the functioning of
the rest of the body. A disharmony of one of the parts causes an
imbalance in the whole. And that is, I believe, the main objective
behind the building of the Great Pyramid. Because the human
being has forgotten his task in life, the whole divine energy field
is out of balance.

We have seen that humanity is moving from the Pisces era to

the Aquarius era, which can be compared to the upgrading one does in school from year one to year two, or at the consciousness levels from sphere 1 to sphere 2. The first phase of the first level of a new sphere is always the red frequency, which is about surviving and perseverance. As we move through the phases we learn different things. Gradually, the world enters the storm in full, even though politicians and other authorities may have you believe that the storm is already behind us. Be well aware that that is not the case!

Let's make use of the help we are being offered to grab hold of our task and restore harmony within the divine field. The designers of the Great Pyramid knew how evolution works and they knew that such a disharmony in the future was on the cards. They took precautions to provide us with a guide to show us how to get through that difficult time.

The day we are confronted with disease, life changes for many of us and the question "why me?" appears in our consciousness. Disease is a hard message about the lesson you are facing in life. The problem is not the lesson, but our stubbornness not to learn it, not to change. We are not responding to the new challenges ahead and so we are missing the point. Imagine you have an exam tomorrow and you are well prepared and feeling confident. The next day you get the exam paper and you notice that you have prepared for the wrong exam. Result: you failed! Now you get the chance to retake the exam in September. Whilst the other students enjoy a well-deserved holiday you now have to study and prepare for the exam. You are being punished because you did not pay enough attention when it mattered.

Let's see what Egyptian mythology, the Bible and the Koran have to say about you taking your exam.

To the ancient Greeks, the heart was the central organ of consciousness. Here, they also located memory, intelligence, feeling and imagination. To them it was the most important organ of the entire body. Everything was collated there. The weight of the heart is determined by the total sum of the light, pure deeds and the heavy, dark ones.

In Egyptian mythology, the Weighing of the Heart was the name of the ceremony a deceased person would undergo in the Hall of Two Truths. Before being allowed to enter into the Kingdom of the Dead they were required to appear before the judge, Osiris, and 42 other judges, where they had to justify their deeds. The heart would then be placed upon one side of the scales whilst on the other side the goddess Maät placed the Feather of Truth. If the heart was lighter than the feather, the deceased would immediately be granted access to the Kingdom of the Dead by Osiris. If the heart was as light as the feather, the deceased moved to the gates of Yaru, the Hereafter. If the heart was heavier than the feather, the scales dropped down on that side and the monster Ammit could then reach the heart and swallow it. Another monster ate the house in which the soul resided and the soul was then destined to roam earth forever.

Here the heart was seen as part of the balancing theme. In Egypt, this balance was held by gods, and was forever challenged by forces that brought chaos. If a person had done a good job in holding the balance, then that was, after his/her death, demonstrated by weighing of the heart by the divine judges. Truthfulness was the norm; lies were taboo. The Egyptians believed that the human being carried on living after death and that his/her deeds were weighed against the feather of Maät. He who appeared before the judges of the death without sin would live on as a god. The feared test, the weighing of the heart, was waiting for all deceased, but it wasn't the only evidence brought to the judgement. The deceased was allowed to plead his/her case in an effort to convince the judges of his/her righteousness. In order to help him/her with this plea, people put texts and sayings on papyrus next to the buried body (The Book of the Dead). The Egyptians were well aware of the fact that very few would appear before the High Judges without sin. So it was imperative that the gods would need to pardon all sins and purify the sinner. The papyrus scrolls, beautifully decorated and scripted texts, were laid between the legs of the deceased. This ensured that the deceased was granted peace and the heart was not eaten by Ammit.

The counterweight was the feather of Maät. Everyone had to be in balance with Maät, whose name meant "the righteousness", "cosmic order" or "truth". Maät thus became the norm to which everyone had to aspire. It wasn't so much the consciousness of the Egyptian as well as the engine that drove this consciousness. Maät was the order within the cosmos, within nature, within everyday life, but also within words, music and art. Maät was also the guardian of the cosmic order, responsible for the cycle of life and the representation of all manifestation of the cosmic laws, the truth and world order. Maät was as old as creation itself.

Doomsday, or Judgement Day, is a concept that returns in Jewish, Christian and Islamic religious books, and refers to the day on which every human being will be judged by God. Doomsday lies at the heart of the eschatology* of these three religions. Doomsday refers to the event whereby God will reward the righteous ones and punish the wrong-doers. Together with death, heaven, hell and purgatory, Judgement Day belongs to the so-called "end-of-time-things".

In the Bible, both in the Old and New testaments, there is mention of an end to time, whereby the survivors and the dead will be judged. In Christian eschatology, Doomsday has an important role to play if we consider the number of times it is mentioned in the Bible. According to John in Revelations, Doomsday happens after a reign of peace that will last a thousand years, and judgement will be held on a great white throne (Revelations 20:11). Then the

---

* Eschatology is "teachings about the end times". This is a Western term referring in Christianity, Jewish teaching and in Islam to all events relating to the end of humanity and the end of time.

dead will arise from their graves to be judged together with the then-living. Satan and all the fallen angels, as well as the people that have spurned the grace of God, will be banned to eternal darkness, where they will be tormented day and night. Those who have embraced Jesus Christ as the saviour of sins and the mediator between God and the human race, will be granted justice in the eyes of God, and they will inherit eternal life beneath a new sky and upon a new earth. God will live amongst them.

In Matthew 12:36–37: *"But I Tell You That Everyone Will Have to Give Account on the Day of Judgment for Every Empty Word They Have Spoken. For by Your Words You Will Be Acquitted, and by Your Words You Will Be Condemned."*

In 2 Peter 3:7: *"But the Heavens and the Earth, Which Are Now, by the Same Word Are Kept in Store, Reserved Unto Fire Against the Day of Judgment and Perdition of Ungodly Men."*

In the Koran there is mention of an end time and the Doomsday (Yawm al-Qīyāmah) crops up several times too. The Hadith Sahieh of Al-Bukhari states broadly what will happen on Doomsday. After death, the souls exist in an intermediate stage, the barzakh. Only when all people and all djinns have died will Doomsday occur. It is then that it will be decided where everyone is placed in the akhirah. On that day, martyrs (shahied) and some sahaba will get their place directly in the divine gardens of Paradise (djannah). The others who have been woken from the grave will be interrogated by the angels, Nakir and Munkar. At the end of that, the deceased will receive a book in which his/her life has been recorded. The righteous will receive it in their right hand and continue straight on into the paradise gardens of the angel Reduan. The wrong-doers will receive the book in their left hand and they will go to hell, djahannam, where the angel Malik rules.

REFERENCES:

Bruyn, Eric de (2001) De Vergeten Beeldentaal van Jheronimus Bosch, 's Hertogenbosch, Heinen ISBN 90-70706-35-0

Recurrent Diseases

It becomes quite clear now how disease can recur in people even after they have received appropriate treatment. Disease is closely linked to evolution. As evolution has a direction, it doesn't make any sense to try to oppose it. Our current treatment protocol is built around the concept of pushing the disease back, and attempting to bring the body back to a previous balance. In principle that could be okay to do, but now we know that you will always have to develop and move forward. By trying to hold the disease back, by trying to regain a previous balance, the person will always, given time, get confronted again with the disease because you can't alter the direction of evolution and you can't stop it.

Resonance

A resonant frequency is a frequency that a system naturally vibrates at when it is triggered away from its balance position. These resonant frequencies are inherent to the system and form the basis on which the various elements that make up matter remain together, even when put under pressure.

Every system has at least one resonant frequency. This is a frequency of vibration that happens in a natural way whenever the system has been moved away from its balance point and has been released. This can happen whenever the system is stimulated to move by an outside force. Harmonic frequencies are sub-notes of the frequency spectrum of the whole. Some objects have more than one frequency. A tuning fork has only one resonant frequency as you always hear the same note whenever it vibrates. A pendulum only has one frequency too, which depends on the length of the cord, not the weight of the mass.

So what does all of this have to do with a person, and, more specifically, with the body of that person? As we are made from frequencies, our interactions with energies happen in a similar fashion. We "resonate" with our environment. We are the total sum of frequencies interacting. We are the resonance frequency of the human frequencies that make us. Let's see if we can create a visual picture of what happens to the body when we respond to energetic influences.

Disease

We represent a person lying on their back in the following fashion:

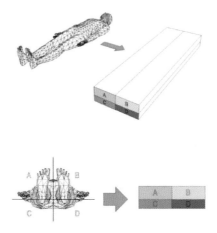

We can represent the body in four quarters, depicted as blocks A, B, C and D. The left side of the body (B and D) is a mirror image of the right side of the body (A and C). But A and B, the front of the body, is not a mirror image of C and D, the back of the body, as the back is more compact than the front.

At any level we can make a sliced picture of the body, which can be represented in the following way:

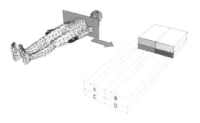

You might wonder how a healing process takes place when disease or disharmony – different words for the same thing – occurs.

Let's begin by creating an ideal situation whereby points c and p have an established harmony.

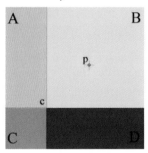

When the ideal point p has moved to p' the harmony has been broken. The vibrational frequency (the resonant frequency) is out of balance. This corresponds to a false note being played! This moves point c to point c', which shows that the entire structure has become imbalanced.

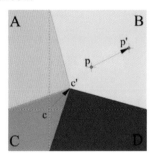

By treating this imbalance correctly, point p' will start to move back in the direction of point p and c' will move towards c. This is moving towards *health*.

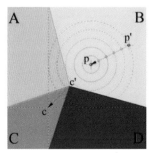

The two points, p and p', represent two different circumstances of life. Situation p' represents a diseased state and situation p represents a healthy state. Of course, we would like to move from p' to p. To get there we must take a number of steps depending on how far we have strayed from our harmony situation. These are represented in the drawing by the concentric circles on our way from p' to p.

By exposing the area to the right resonant frequency, something is being activated; you start your journey. You become aware of the background noise of your situation. Step by step you will find answers to the question "Why me?". It doesn't come easily though; you will have to work for it.

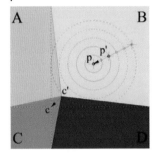

The next drawing shows how point p' has been moved in the direction of p, in spite of it still being somewhat removed from it. Because of the movement of p' your circumstances will also change. This happens because you have responded to the opportunities life has offered you, and these, in turn, are related to the lesson you need to learn. You will be confronted with the "mistakes" you have made with regards to the lesson. The return path from p' to p is the opposite direction you have gone through in order to end up at p'. Therefore, you will have to endure it again! This does not necessarily mean that you will ever remember spending time in situation p. The only thing you know is that you have kept moving away from p. Point p is the perfect situation for you to learn your lesson.

The better use you make of the opportunities on offer, the closer you will end up to p. Remember the formula $Ep = Mp.c^2$? When the energy pattern changes, the mass pattern changes too. The only thing is, it may take some time before it shows, but it will move to point c too.

Never forget that this is a process and the time it takes depends on how far the return journey is and how fast you move. In other words, it depends on how well you make use of the opportunities offered to you. The last few steps often turn out to be the hardest of them all. You will have to swallow 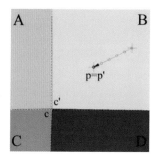 the last morsel. When we learn something new it takes all our concentration and it is a huge effort. Changing attitudes, changing reaction patterns, is no different, but gradually you realise you are no longer consciously giving it any attention because the new way has become "normal".

Point c' will, in reality, never fully match point c. If that were the case then no more movement would be possible or needed, as perfect harmony would exist in life. However, in our real lives there is no perfect silence, no perfect quietness. There is always movement. The movement happens around the point of balance, the point of stillness. This is the power that drives evolution. This is why everything changes all the time and nothing stays the same. When your life reaches harmony, you experience heaven on earth.

Let's take an ordinary radio set as an example. The radio is tuned in to Radio Ideal (point c). The songs broadcast by this station are your favourite tunes, presented by your favourite DJs. This station makes you deliriously happy. Here you meet yourself constantly. Here you can experience your lessons in full. It is, for you, the ideal place to be.

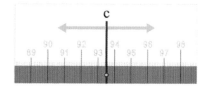

Now imagine that your life's lesson is situated in the red zone (red in A of the spectrum of visible light). It is about surviving. Your birth takes place in circumstances that will allow for this lesson to be learned. If your life's lesson is about survival, the chances are that you won't be a child of Bill Gates' household! It makes a lot more sense to place such a human in a dangerous ghetto situation, for example. So, the environment provides the circumstances in which the lesson can be learned.

The problem is that we very rapidly lose touch with why we came here. Modern society has also become so complex that the potential is almost endless in relation to the kind of lessons to be learned. This, of course, also means that our path is strewed with possible escape routes. We make use of everything to provide "happiness", to avoid having to face up to the lesson. We lose contact with the real path more and more, until one day it happens: you don't feel so well and it isn't going away. You get help, investigations are performed, and it is confirmed – you are ill. It would be better to not say you *Are* ill, as you do not need to identify yourself with the illness. It is more that **you have an illness**. Now you have the possibility to release yourself from it.

The drawings below illustrate the different environments away from the ideal point, illustrated by point c'. This can happen to the right as well as to the left. We could say that in one situation there is too much c and in the other too little c.

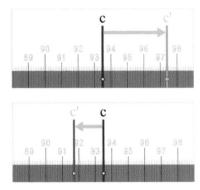

By turning the dialling knob, you move from frequency c′ towards frequency c, your ideal frequency. On the way you meet other stations that might distract you from where you want to go (c″, c‴, etc.). You need to let go of a lot of things you were holding on to in order to focus on what is truly important in your life. This brings you closer and closer to point c.

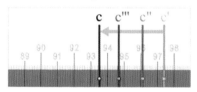

Once you get in the neighbourhood of c, you are confronted with the reality of your life's lesson. You feel a harmony entering your life. You no longer have the illness. You are tuned back into Radio Ideal. It will still remain a battle and you still have to be alert. Just as it was at school, you may get confronted with an unexpected challenge. You will also have to stay focussed and revise regularly. So, in life you will have to remain alert and focussed and stay in touch with the priorities of your life. You need to be present in the "now" all the time.

Disease or Disharmony

From what we already know, we can deduce that disease/disharmony can only occur when there is a constant dissonant frequency that we allow to be present. This moves the frequency of the whole away from its balance point. There are only two possible directions in which this can happen: either the frequency gets elevated (yang-effect in kinetic energy and yin-effect in mass energy) or the frequency gets depressed (yin-effect in kinetic energy and a yang-effect in mass energy). As human beings, we have not evolved enough in order to read energy fields easily and so it may be wise to focus on reading the material effects primarily.

When tissues become harder, or more compact, it is a yin effect.

This means, in general terms, that the pressure on the person, and his/her tissues, is too high. So, the end result of all impacts is an increased pressure over a long period of time. This is the reality for that person at that time; this is the way it is. Any thoughts on the efforts I have already made or all the "positive" changes I have introduced into my life are a waste of time, judged by the reality of my tissues, by "the state I am in".

When the tissues are loose, or weak, they show a yang effect. This means that there isn't enough pressure on the system and so everything runs slowly and ineffectively. There is a lack of power, a lack of energy, running through the tissues. We lack motivation and execution power.

Essentially these are the only two disease patterns. And all of the extensive medical literature about separate illnesses can be put into one of these two categories. From an energy point of view, there are only two possible deviations from the balance point and all manifestations of this can be brought back down to one of these two directions. What happens to the tissues when the frequency changes?

Tissues are only exposed to two influences and forces: the inner force of an individual – a centrifugal force – and the outer force from the environment – a centripetal force. The balance between the inside-out force and the outside-in force gives us the tension within the tissues. The tissues have been formed in a very specific way, with very specific information, and have thus been prepared to absorb a specific set of pressures. This is the result of the powers that created and formed the foetus and the baby, which makes all of us unique. When the balance between the two existing forces changes, the tension within the tissues changes too. This means that just by growing up we will have a different relationship with the environment, and hence the tension within the tissues will change during the course of our lifetime.

A disease process is an ongoing pressure change between the two forces that run through the tissues. It has to be sustained for a period of time, which means that no disease attacks us out of the blue or kills us in an instant. It is only slowly that the tissues are forced to change function and then shape as a result of the pressure. The only way they can respond is by making the tissues either more compact – harder – or less dense – looser. This leads to two types of effects: a yin effect and a yang effect.

A yin effect makes the tissues harder, denser, and this can occur in two ways: either the outside force is greater than normal, or the inner force is smaller than normal. Under both conditions the tissues become squashed. An increased pressure on the outside occurs when the environment is demanding more of the individual than he/she can withstand. The other possibility is when the inner force is being oppressed and some talents of the individual are being opposed. This increased outer pressure over a period of time will start to create an increased pressure within the tissues, which, in time, will become "permanent".

On the other hand, it is possible that the inner force has diminished, either through a lack of will to do something with this life or through enormous fear that paralyses the flow. This gives the impression that the forces on the outside have gained strength. Life is made up from tension. Evolution happens because of tension. It is therefore not a requirement that we get rid of all the tension. All we need to do is to keep the balance of all occurring tension. There will always be outside pressures and we need to counterbalance that with inner power. In order to survive we need to **want** to survive!

A yang disease pattern happens because the tissues become less dense, and here the cause lies in either a lesser outside force or a greater inner force. When the environment isn't strict enough, or not demanding enough, then the individual will quickly outgrow himself or herself, which doesn't give the tissues enough time and resistance to gain strength. The person does not form properly, they lack the inner power to maintain the growth process. We find examples in individuals who have never been opposed, have never been denied, have never had to fight for something. They have never had to "prove" themselves.

We find the same end result of becoming overstretched when

the inner power of the individual is so strong that it almost destroys the environment; it blows it away. These are people who are almost unstoppable in their behaviour, which seriously splits their surroundings into two camps: one will flee and the other will adore and bask in their radiating glory. This is like living in a pressure cooker and these people tend to burn themselves up very quickly.

All hardening processes can be viewed as yin disease patterns. We find examples in stiff painful muscles, in arthritic conditions of the joints, in hardening of the nervous tissue (MS, Parkinsonism, dementia), in hardening of blood vessels, and, of course, in the ultimate hardening process: cancer. In this category we also find hardening processes of the mind, such as depression, extreme stubbornness and listlessness.

All loosening processes can be viewed as yang disease patterns. This becomes clear in all processes that turn solid tissues into fluids, as in the formation of pus; the debris of cellular disintegration. This means that inflammation and infection belongs in this category. *Inflammation Means the Presence of All Four Symptoms: Pain, Swelling, Redness and Warmth. A Diagnosis of Inflammation When One of These Symptoms Is Missing Is Simply Wrong!* All experiences of burning sensations or fever are yang developments. In the mind we encounter signals, such as restlessness, hallucinations, fantasies, extreme irritation and an explosive temperament.

As you can now see, it makes a lot more sense to describe illnesses in terms of their signs and symptoms than in terms of a medical label. We need to understand the forces that cause the reactions. In the West we have accepted naming diseases in a way that removes all identification with the underlying forces, which has allowed the diseases to become divided into very small pieces, each with their own specialist. No attention is being paid to the whole anymore. The question: "Is it expanding or contracting?" does not surface anywhere.

Once we have decided which disease pattern we are facing, then all we need to do is to determine the underlying cause of the imbalance. What follows is all you ever need to know about the causes of disease.

- A yin disease is caused by either a larger than normal outer pressure on the individual, or a smaller than normal inner pressure coming from the individual.
- A yang disease is caused by either a larger than normal inner pressure coming from the individual, or a smaller than normal outer pressure on the individual.

(Normal = the balance point between those two forces during the creation of the individual.)

Now we know this, how are we going to treat diseases effectively?

1. Yin – larger outer force:
This can be treated in two ways: either the outer force diminishes (because the individual has asked his/her environment and has been heard), or the individual has to remove himself from the environment.

2. Yin – smaller inner force:
As the outer force is as expected, it does not stand to reason that this will change simply because one person would like the world to be different. Unless the individual finds himself a completely different world, such as a totally different society, then the necessary change to restore health is not going to come from the outside. For most people, the only solution is "to toughen up". It is imperative that they increase their inner strength.

3. Yang – larger inner force:
First of all, you need to realise that it is you who is burning yourself and your environment up. Being aware of this allows you to make different choices, such as creating space for rest and quiet time, to "cool it". This applies to not just the physical side of life, but also to the mental aspects. In an early stage of this development, the environment still has an opportunity to exert some control over

the individual, and to confront him or her with the effects of their behaviour so that together a better, more sustainable balance can be found.

4. Yang – smaller outer force:

When the inner force is truly not exaggerated then it is up to the outside world to "toughen up", and to take measures to change this effect within the individual. However, this is unlikely to happen because the surroundings of the individual haven't acted before. We can only hope that seeing the individual becoming ill might be the trigger for the awareness and the subsequent action to oppose the individual. If this does not happen then the individual must go and look for a more restrictive and oppressive environment. "Challenging" is then the buzz word!

Maybe you have already noticed that in order to ensure a path to health, the answers seem to lie with the individual. It is individuals who decide what needs to change and how, either by changing their environment or by changing themselves. When the environment is incapable of changing, or unwilling to, then we are back to the individual who can decide to make the necessary changes. Being in balance, being healthy, becoming healthy, depends entirely on the decisions the individual makes. When you begin to make the changes, you are on your way to restoring health. If you don't make the necessary changes and you hold on to the same patterns that got you to this point, the point of disease, then you remain on the path of disease. And no amount of therapy, advice or outside help will cure you. The only one who can do that is *You*. The only contribution any therapy can bring to your healing process is to relieve the suffering for a while, but not to change the process. When you know what they can and cannot do, you can make effective use of them, as and when required. Just be aware of attaching any unfounded expectations to them.

There is also a difference in how you make use of therapeutic methods. Using therapies to suppress the signs of the true status of the system only leads to a reaction from the system against the oppression. The system needs to bring a certain message to your consciousness, and it will keep on trying, in spite of your suppression. In this way the system generates an enormous amount of power against the oppression by energising this particular fight, instead of energising the healing process. Suppress the pain and the pain impulse will get stronger and will eventually surface again. Suppress the inflammation and the system will create more heat to increase the yang effect it is seeking, which will result in chronic inflammatory diseases, or it will dispatch the problem to another area where it has more room to break through. Examples of this are recurrent throat infections or throat infections that eventually become a lung infection. Suppress the blood pressure and the system will increase the internal pressure in order to be able to operate, for which it requires a higher than normal pressure.

When you realise that despite whatever method or product you use as a therapy, all you are doing is sending a message to your system to try to elicit a response, and you will use these differently. If what I need to create within my tissues is a yang effect, then it doesn't make sense to oppose the pressure that is already there. Remember that the tissues react to the message, which means that they create an opposite effect than the immediate effect of the message. By increasing the outside pressure, you evoke an even larger pressure from the inside. Then you remove the extra outer pressure and the sum total has resulted in a yang effect because the inner pressure is now larger than the outer. For example, in order to relax muscles, to reduce the tension inside them, you pressurise the painful tensed muscles, which will create an even larger pressure inside the muscles. When you then remove the pressure you have put upon them, you are left with a larger inner force moving outwardly, thereby creating a yang effect within the muscles, giving them more space and allowing them to relax. In order to remove pain, you increase the pain. In order to remove the disease, you aggravate the disease. Isn't that a basic homeopathic principle?

This healing method – the only one – only works when the

individual has enough energy, or enough inner strength, to react to the incoming messages. Once that power becomes insufficient to complete the task – healing is one – then the system can only resist for as long as it can. There does not exist a healing method for an empty vessel! These are the rules of life and it is about time we learnt to know and accept them. We are only here for a short period of time and for a specific reason. Let's not try to hold on to something that isn't meant for us or for longer than we are allowed. We do not own life, so let's stop trying to make it ours. We will only succeed in making ourselves very unhappy by not accepting it as it is.

Now that we have decided which disease process it is, it makes sense to take a closer look behind the scenes. Imagine a system that has been pressurised for a very long time. It will respond automatically without waiting for permission. Yin disease leads to a yang response, so you will encounter inflammation and infection as the system tries to open itself up. This yang effect will not be sustained, but shows how the system struggles to overcome the effects of the ongoing pressure. So, although all you see and experience is a yang effect, this is not the disease process. That goes in the opposite direction! This means that if we want to help this system we need to be looking to support the efforts to create a yang effect, not oppose it. It would then be a good idea to encourage inflammatory processes and infections to take place, whilst we start working on a plan about reducing the ongoing outside pressure on the system.

Never forget: **The sign is not the disease.**

It boils down to the fact that the resonant frequency of our system, the way we have been put together, has been disturbed by interference and influences of different frequencies. The only way to restore the resonant frequency is to move in the opposite direction from the one that created the disturbance in the first place. And that is your lesson in life, to do that. Once you manage it there is another challenge waiting for you. In effect, it is trying your best, and trying again until you succeed, and then keeping on trying to improve on it until you manage it routinely. Then it will feel as if you have always known it. That is how we learn everything in life, from standing up to walking to running to cycling. First, you work hard at it and you need a lot of help. Later, you can do it by yourself, and later still you do it effortlessly, or so it seems. Only now and again a mishap may occur and you need to refocus; you need to stay in tune with the lesson.

A "helping hand" is what the correct treatment is. You need a guideline that tells you how to get from disharmony to harmony. From our research it has become clear that within a few years it will be possible to make that move, not via a physical route as it is attempted now, but by using resonant frequencies on the human energetic system. We already know that when we alter the human energetic field we will change the physical expression, the body, too. By using specific resonant frequencies on the system, the person is going to become conscious, by way of the corresponding expressions within the body, of lessons to be learned.

In life, you come across many crossroads, which face you with the challenge of making decisions. These decisions result in experiences, which will show themselves to you as being "right" or being "mistakes". All these together create the picture of what is "right" for you and the only thing you need to do is to keep trying. Eventually you will get closer and closer to the *Real You*, and once you have discovered your basic tuning you have successfully completed the lesson. Now you move on to the next bit of learning. The whole thing is not easy and on many occasions you feel like quitting, as the challenges you are faced with are hard and sometimes look impossible for you to learn.

So we can conclude that when life has been out of sync for some time, we will receive signals of disharmony, which we call the disease process. This will eventually lead to a malfunctioning physical and mental structure, which is going to need a bit of help to restore itself. To make the journey back to health we need messages that point the system in the right direction; the direction of its resonant frequency. The natural system will then automatically follow that direction and, as long as the system can generate enough energy, it will keep on moving towards its balance.

In the future, we are going to be able to put the message into the energy field itself by introducing the right note in the right direction in order to create an interference pattern of the resonant frequency. That way, the system will "pick up" the input frequency and respond by creating its own resonant frequency.

The entire universe is on the move as a result of two antagonistic but complementary forces: yin and yang. These are terms we have adopted from the Chinese but we have to be careful not to see them as absolute; they are a measure of relativity. They are dynamic and compare directions of movement in relation to other things. There is no such thing as absolute immobility.

Yang is the result of a centrifugal force, moving away from the centre. This is the force of expansion, separation and enlargement.

Yin, on the other hand, is caused by a centripetal force, moving towards the centre. This is the force of contraction, togetherness and shrinking.

Now we can see how this interacts within the seven different layers of the light spectrum. Here is a reminder of the proportional division of the colours within the spectrum:

7	3,56707 %
6	15,11025 %
5	5,77156 %
4	24,44888 %
3	2,20455 %
2	9,33863 %
1	39,55906 %
	100 %

This is the same proportional division that we encounter all the time, but the colour combinations are different each time. On each occasion we can calculate the proportional representation of each of the colours.

The seven chakras show up in 343 horizontal planes across the human body. Each chakra shows up in 49 of these planes. The different composition of each of the chakras in these 343 layers can be calculated based on the known percentages. Then we can determine the yin–yang configuration of each of those, in the same way we have done before. This will allow us to compare and to seek connections between layers and frequencies. You can even calculate the resulting effect on each chakra of a centripetal or centrifugal resonant direction. It is fascinating!

Chapter 11
The Human Energetic Circuit

When confronted with a disharmony in life, it is possible to locate the disharmony within the body. Its place can be related to the divided spectrum of visible light, which leads us to identifying the chakra that is out of balance (part c in the division of the spectrum). We also note whether the disharmony manifests on the right or the left side, and on the front or the back.

We have already mentioned that within a few years a disharmony (disease) won't be treated physically anymore. Instead, the human energetic system will be brought back to harmony by using resonant frequencies. From our lengthy research we know that there are seven resonant frequencies that can be used on each chakra.

The following picture represents the energetic circuit:

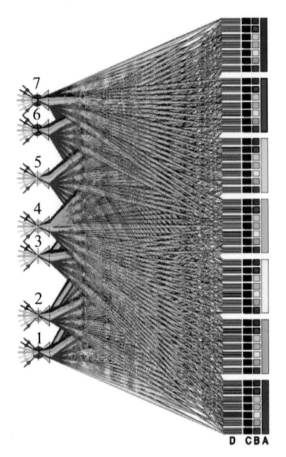

D CBA

- On the left we have the seven chakras, each with the seven
  resonant frequencies.
- The chakras are represented as prisms, like the ones that break
  light into colours.
- On the right we see the diversification: A for the visible spectrum; B for the first colour division; C for the areas (343 in total)
  where the chakras show themselves; and D for the composition
  of the various systems.

The result is a complex drawing that is difficult to read. However, if we compose the drawing in layers per chakra, it brings a bit more clarity to the multitude of lines.

Let's just look at chakra 1:

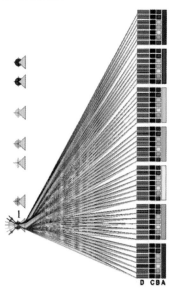

To the left we see the first chakra with seven resonant frequencies. The chakra opens up the frequency in a way a prism opens up visible light.

To the right we have the codes: A for the visible light spectrum; B for the human energy field; C for the individual energy field with its chakras; and D for the composition of the various systems.

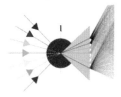

This diagram shows detail of chakra 1 with the seven resonant frequencies:

Now, let us simply draw resonant frequency 1 on chakra 1:

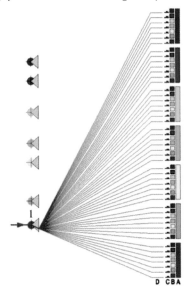

We notice that resonant frequency 1 manifests in 49 areas where chakra 1 is active. Does this allow us to conclude that the various resonant frequencies can be directly linked to the various systems we have in our body, such as the lymphatic system (1), the circulation system (4), the respiratory and digestive system (6), the motor system (2), the sensory system (5), the nervous system (7) and the excretion and glandular system (3)? Not that straightforward!

Here follows a correction: In Chapter 10 we linked the human evolutionary system to our schooling system and we used a certain coding system to go with it. The codes referred to the various lessons to be learned during the evolutionary process.

Alternative View on the Human Energetic Circuit

However, it turns out that my drawing of the energetic circuit of the human being was wrong and needed an adjustment. I apologise for that but we only learn as we go along! Here is the adjusted picture:

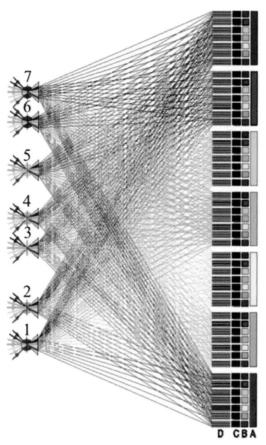

- On the left we have the chakras, each with seven resonant frequencies.
- The chakras are drawn as a combination of a circle and a prism. (Think about the light breaking into seven colours.)

- On the right we have code A – the spectrum of the visible light;
  code B – the human field; code C – the individual field with
  the sites of the crossovers of the chakras (343 layers); and code
  D – the composition of the various primary systems.

When we isolate chakra 1 in this drawing, we get the following:

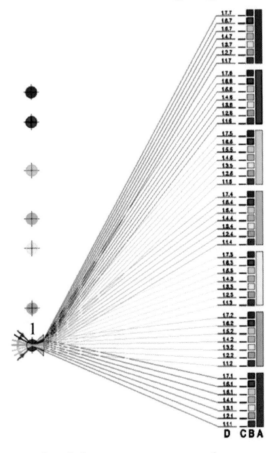

In the next detailed picture, we can see that every resonant
frequency within the prism is split into seven lines or seven separate circuits:

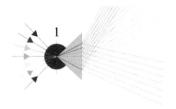

When we now filter out resonant frequency 1, we get the following picture:

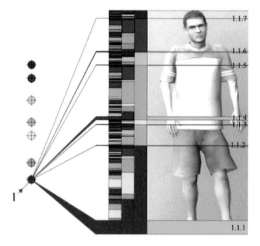

The actual circuit drawing for resonant frequency 1 on chakra 1 looks like this:

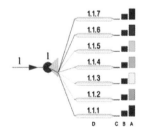

Now we have resonant frequency 1 influencing only seven areas. These are, using the code system we created for the human evolutionary system: code 1.1.1, 1.1.2, 1.1.3, 1.1.4, 1.1.5, 1.1.6, and 1.1.7.

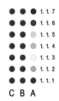

From Chapter 10, we know the following:

- In 1.1.1 the bold figure represents the type of education (pre-school, junior, college, etc.). We find this information in column C.
- In 1.1.1 the bold figure represents the class, which we find in B.
- In 1.1.1 the bold figure represents the subject, which we find in A.

Now we can see that all the subjects of one class year are highlighted by one circuit.

Resonant frequency 2 on chakra 1 gives us the following picture:

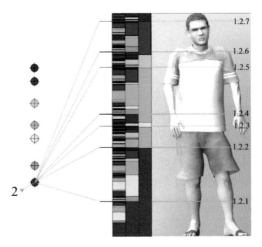

The circuit for resonant frequency 2 on chakra 1 is as follows:

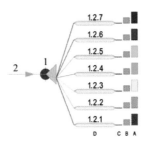

The corresponding codes are: 1.2.1, 1.2.2, 1.2.3, 1.2.4, 1.2.5, 1.2.6, and 1.2.7.

Resonant frequency 3 on chakra 1 gives us the following picture:

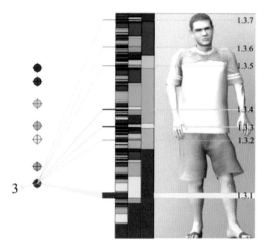

The circuit for resonant frequency 3 on chakra 1 is as follows:

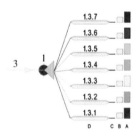

The corresponding codes are: 1.3.1, 1.3.2, 1.3.3, 1.3.4, 1.3.5, 1.3.6, and 1.3.7.

The resonant frequency 4 on chakra 1 gives us the following picture:

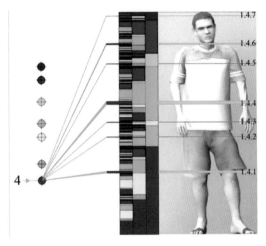

The circuit for resonant frequency 4 on chakra 1 is as follows:

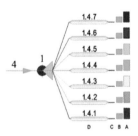

The corresponding codes are: 1.4.1, 1.4.2, 1.4.3, 1.4.4, 1.4.5, 1.4.6,
and 1.4.7.

The resonant frequency 5 on chakra 1 is as follows:

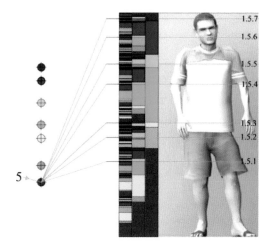

The circuit for resonant frequency 5 on chakra 1 is as follows:

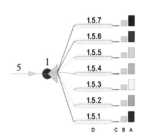

The corresponding codes are: 1.5.1, 1.5.2, 1.5.3, 1.5.4, 1.5.5, 1.5.6, and 1.5.7.

Resonant frequency 6 on chakra 1 is as follows:

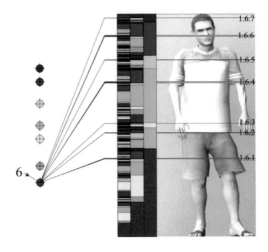

The circuit for resonant frequency 6 on chakra 1 is as follows:

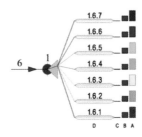

The corresponding codes are: 1.6.1, 1.6.2, 1.6.3, 1.6.4, 1.6.5, 1.6.6, and 1.6.7.

Resonant frequency 7 on chakra 1 is as follows:

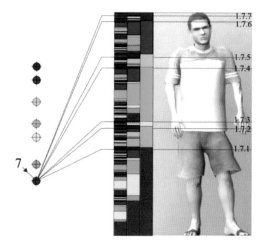

The circuit for resonant frequency 7 on chakra 1 is as follows:

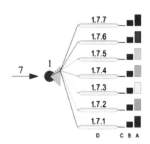

The corresponding codes are: 1.7.1, 1.7.2, 1.7.3, 1.7.4, 1.7.5, 1.7.6, and 1.7.7.

In this way we can draw every single circuit per chakra and per resonant frequency. Take resonant frequency 2 on chakra 2 as an example. This is the picture:

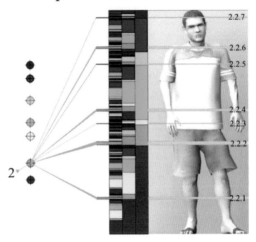

And the circuit:

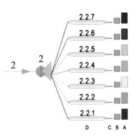

And resonant frequency 4 on chakra 2 gives the following picture:

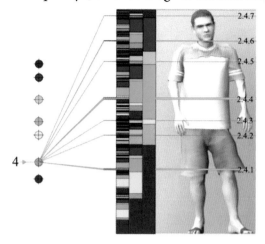

The circuit:

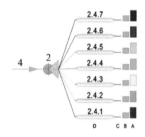

Resonant frequency 5 on chakra 4 gives us this picture:

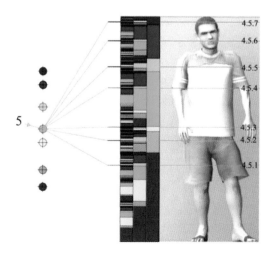

Finally, we can truly comprehend the division of the visible light spectrum. And that in turn makes it possible for the human energetic field to be analysed. We can now truly relate the disharmony in a specific place on the body to the frequency we require to place on the specific chakra in order to restore harmony. It is the colour in column B that determines the resonant frequency. The colour in C determines the chakra. This gives us information about which resonant frequency is needed on which chakra to rectify a disharmony. The drawings show us that every resonant frequency on a chakra influences seven areas of the body.

So, what to do next with this information?

By dividing the visible spectrum we obtained, in column C, 343 layers. This needs to be multiplied by four because the body has four separate parts: left front, left back, right front, right back. Now we have 1,372 separate parts. From the above analysis we know that every resonant frequency operates seven parts, which means that we need to set up 196 ways to cover all the possible resonant frequencies. Let's try to prove this!

From our tests we found that when we used a resonant frequency on the left side of the chakra it "lit up" areas in the left side of the body as well as the right side, which was also the case when using a resonant frequency on the right side of the chakra. Even more interestingly, the same was true for the front and the back of the body, which are both touched by frequencies on the left and the right side of the chakras. Let's illustrate this in some drawings.

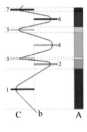 The next diagram shows the pathway of an energy wave, b, throughout the visible spectrum A. In C we can see the seven different parts that get affected by the energy wave b. The red layer (1 left) represents the layer that has been activated in the red part of the visible spectrum. The orange layer (2 right) represents the orange part activated within the visible spectrum, and so on.

The next drawing shows the paths of two separate energy waves throughout the visible spectrum. The separate paths can be coded too. The blue path is coded 1–3–5–7 right and 2–4–6 left. This means that the red, yellow, blue and violet parts will be activated on the right, and the orange, green and indigo parts will be activated on the left. For the red path the code is 1–3–5–7 left, and 2–4–6 right, meaning that the red, yellow, blue and violet parts will be activated on the left and the orange, green and indigo parts on the right.

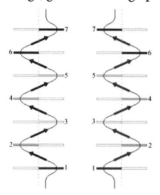

If we combine the two pathways then we have:

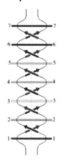

When we enlarge the previous drawing, looking from behind so that the left side of the body shows up on the left and the right side shows up on the right, we notice that on the switch from layer 1 red to layer 2 orange, *The* BLUE *Path Crosses at the Back of the* RED *Path,* and on the switch from layer 2 orange to layer 3 yellow, RED *Crosses at the Back of the* BLUE.

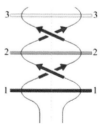

We can visualise this better in a three-dimensional picture, which also allows us to see that the energetic pathways spiral upward.

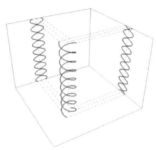

Our entire body is a materialisation that occurs via seven chakras, or seven energy portals. The body itself has four identifiable areas: left back, left front, right back and right front. We know that, per chakra, we can use seven resonant frequencies that can be specified in such a way that every energetic line, connected to the specific chakra, can be harmonised. In total there are 196 energy lines.

Let's visualise these lines. There are four possibilities.

Possibility 1:

The left drawing below is a 3D image. The right drawing is a representation of the energy lines within the various components of the visible light spectrum.

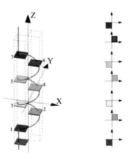

Possibility 1 shows us the different places within the visible spectrum where these energy lines become matter:

- Left back in the red range (1)
- Right front in the orange range (2)
- Left back in the yellow range (3)
- Right front in the green range (4)
- Left back in the blue range (5)
- Right front in the indigo range (6)
- Left back in the violet range (7)

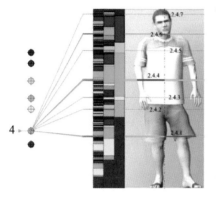

The next drawing clarifies this further. We take, as an example, energy line 4 of chakra 2.

In areas 1, 3, 5 and 7 this energy line activates the corresponding layers 2.4.1, 2.4.3, 2.4.5 and 2.4.7 left back, and in areas 2, 4 and 6 the layers 2.4.2, 2.4.4 and 2.4.6 are being activated at the right front.

Possibility 2:

The following diagrams show us the different places within the visible spectrum where these energy lines become matter:

- Right front in the red range (1)
- Left back in the orange range (2)
- Right front in the yellow range (3)
- Left back in the green range (4)
- Right front in the blue range (5)
- Left back in the indigo range (6)
- Right front in the violet range (7)

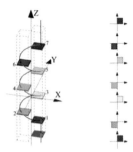

As an example, we take the energy line 4 of chakra 4.

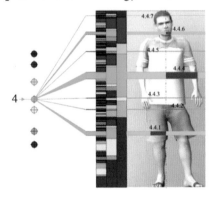

In areas 1, 3, 5 and 7 this energy line activates the corresponding layers 4.4.1, 4.4.3, 4.4.5 and 4.4.7 on the right front, and in areas 2, 4 and 6 it activates layers 4.4.2, 4.4.4 and 4.4.6 on the left back.

Possibility 3:

The following diagrams show us the different places within the visible spectrum where these energy lines become matter:

- Right back in the red range (1)
- Left front in the orange range (2)
- Right back in the yellow range (3)
- Left front in the green range (4)
- Right back in the blue range (5)
- Left front in the indigo range (6)
- Right back in the violet range (7)

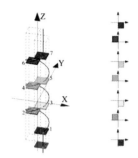

As an example, we take the energy line 4 of chakra 1.

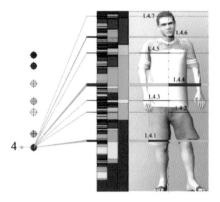

In areas 1, 3, 5 and 7 this energy line activates the corresponding layers 1.4.1, 1.4.3, 1.4.5 and 1.4.7 on the right back, and in areas 2, 4 and 6 it activates layers 1.4.2, 1.4.4 and 1.4.6 on the left front.

Possibility 4:

The following diagrams show us the different places within the visible spectrum where these energy lines become matter:

- Left front in the red range (1)
- Right back in the orange range (2)
- Left front in the yellow range (3)
- Right back in the green range (4)
- Left front in the blue range (5)
- Right back in the indigo range (6)
- Left front in the violet range (7)

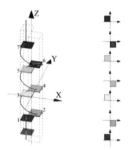

As an example, we take the energy line 4 of chakra 5.

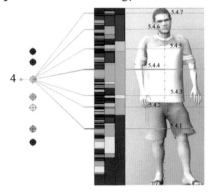

In areas 1, 3, 5 and 7 this energy line activates the corresponding layers 5.4.1, 5.4.3, 5.4.5 and 5.4.7 on the left front, and in areas 2, 4 and 6 it activates layers 5.4.2, 5.4.4 and 5.4.6 on the right back.

From these examples – and this was proven during our testing sessions – we notice that left back and right front, as well as right back and left front, are interchangeable.

The next step is to examine how we can reach every single energy line within the chakra. We already know that a chakra gateway is between parts of the human energy field and the physical manifestation of that field. From our investigations, we can already conclude that a chakra is not one gateway, but a combination of several. In order to get a better picture, we will visualise the chakra as a sphere, consisting of eight separate areas. These relate to our knowledge of the division of visible light, whereby indigo gives us two possibilities (upper and lower) and orange four possibilities (left front, left back, right back and right front). Together this makes eight separate areas.

The sphere with its eight areas has four on the top, coloured red, and four on the bottom, coloured blue.

- Left front upper
- Left front lower
- Left back upper
- Left back lower
- Right front upper
- Right front lower
- Right back upper
- Right back lower

The next picture clarifies this a bit more. On the left is a representation of a chakra in a Cartesian coordinate system. On the right, we pulled the eight separate areas apart to provide you with a clearer view:

Let's start with the four areas on the left side. We give each of these areas a code, based on the combination possibilities of c and f, thereby naming the blocks within the Cartesian coordinate system. What are these possibilities?

Combinations of C and F

By combining C and F we create four possibilities: CC, CF, FF and FC. These activate the various possibilities on the *Left Side* of the chakra.

Upper and lower part of the left back:

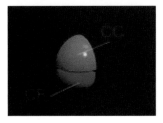

Suppose CC is the setting required to open the upper left back (red) and CF the setting to open the lower left back (blue). A *Setting at the Left Back Side* can direct the energy to flow in two directions. Depending on the chakra and the resonant frequency this gives us the following:

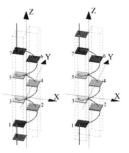

The picture on the left shows us the gate CF on the lower part of the system, and the various activated layers within the visual spectrum. Also note the spiral movement as every sequential layer gets activated. The activated layers **1, 3, 5 and 7** lie in the **left back** area, and the activated layers **2, 4 and 6** lie in the **right front** area.

The picture on the right shows us the gate CC on the upper part of the system, and the various activated layers within the visual spectrum. Here, the activated layers **1, 3, 5 and 7** lie in the **left back** area and the activated layers **2, 4 and 6** lie in the **right front** area.

Hence, depending on the chakra and the resonant frequency, these are two examples of a result instigated by a setting on the left back of a chakra.

However, there is more! Depending on the chakra and the resonant frequency, there are two more possibilites.

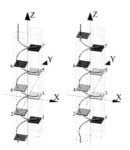

To the right we have the picture of those two other possibilites:

The picture on the left shows us the gate CF on the lower part of the system, and the various activated layers within the visual spectrum. Also note the spiral movement as every sequential layer gets activated. The activated layers **1, 3, 5 and 7** lie in the **right front** area, and the activated layers **2, 4 and 6** lie in the **left back** area.

The picture on the right shows us the gate CC on the upper part of the system, and the various activated layers within the visual spectrum. Here, the activated layers **1, 3, 5 and 7** lie in the **right front** area and the activated layers **2, 4 and 6** lie in the **left back** area.

These activated areas lie on the opposite sides from the first setting.

To Summarize:

The activated areas are the following:

    For layers 1, 3, 5 and 7, activation happens left back, and for layers 2, 4 and 6 it happens right front.

    For layers 1, 3, 5 and 7, activation happens right front, and for layers 2, 4 and 6 it happens left back.

    Upper and lower part of the left front:

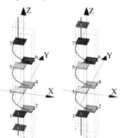

Suppose FF is the setting required to open the upper left front (red), and FC the setting to open the lower left front (blue). *A Setting at the Left Front Side* can direct the energy to flow in two directions. Depending on the chakra and the resonant frequency this gives us the following diagrams:

The picture on the left shows us the gate FC on the lower part of the system, and the various activated layers within the visual spectrum. Also note the spiral movement as every sequential layer gets activated. The activated layers **1, 3, 5 and 7** lie in the **left front** area, and the activated layers **2, 4 and 6** lie in the **right back** area.

The picture on the right shows us the gate FF on the upper part of the system, and the various activated layers within the visual spectrum. Here, the activated layers **1, 3, 5 and 7** lie in the **left front** area, and the activated layers **2, 4 and 6** lie in the **right back** area.

Hence, depending on the chakra and the resonant frequency, we have two possible ways of hitting the left front of a chakra. This leaves us with two more possibilities with potentially different results.

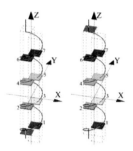

   The picture on the left shows us the gate FC on the lower part of the system, and the various activated layers within the visual spectrum. Also note the spiral movement as every sequential layer gets activated. The activated layers **1, 3, 5 and 7** lie in the **right back** area, and the activated layers **2, 4 and 6** lie in the **left front** area.

   The picture on the right shows us the gate FF on the upper part of the system, and the various activated layers within the visual spectrum. Here, the activated layers **1, 3, 5 and 7** lie in the **right back** area and the activated layers **2, 4 and 6** lie in the **left front** area.

To Summarize:

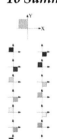

The activated areas are the following:

   For layers 1, 3, 5 and 7, activation happens left front, and for layers 2, 4 and 6 it happens right back.

   For layers 1, 3, 5 and 7, activation happens right back, and for layers 2, 4 and 6 it happens left front.

Combinations of c' and f'

The areas on the right side are also given a code, and this is based on combining c' and f'. This results in four combinations: c'c', c'f', f'f' and f'c'. These will activate the various possibilities on the *right side* of a chakra.

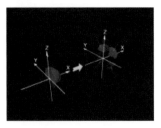

Upper and lower part of the right back:

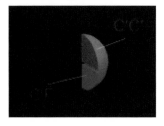

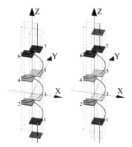

Suppose c'c' is the setting required to open the upper right back (red) and c'f' the setting to open the lower right back (blue). *A Setting at the Right Backside* can direct the energy to flow in two directions. Depending on the chakra and the resonant frequency this gives us the following:

The picture on the left shows us the gate c'f' on the lower part of the system, and the various activated layers within the visual spectrum. Also note the spiral movement as every sequential layer gets activated. The activated layers 1, 3, 5 and 7 lie in the **right back** area, and the activated layers 2, 4 and 6 lie in the **left front** area.

The picture on the right shows us the gate c'c' on the upper part of the system, and the various activated layers within the visual spectrum. Here, the activated layers 1, 3, 5 and 7 lie in the **right back** area and the activated layers 2, 4 and 6 lie in the **left front** area.

There are two more possibilities:

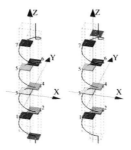

The picture on the left shows us the gate c'f' on the lower part of the system, and the various activated layers within the visual spectrum. Also note the spiral movement as every sequential layer gets activated. The activated layers 1, 3, 5 and 7 lie in the **left front** area, and the activated layers 2, 4 and 6 lie in the **right back** area.

The picture on the right shows us the gate c'c' on the upper part of the system, and the various activated layers within the visual spectrum. Here, the activated layers 1, 3, 5 and 7 lie in the **left front** area, and the activated layers 2, 4 and 6 lie in the **right back** area.

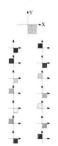

To Summarize:

The activated areas are the following:

For layers 1, 3, 5 and 7, activation happens right back, and for layers 2, 4 and 6 it happens left front.

For layers 1, 3, 5 and 7, activation happens left front, and for layers 2, 4 and 6 it happens right back.

Upper and lower part of the right front:

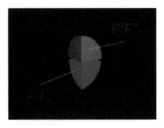

Suppose ғ′ғ′ is the setting required to open the upper right front (red) and ғ′c′ the setting to open the lower right front (blue). A *Setting at the Right Front Side* can direct the energy to flow in two directions. Depending on the chakra and the resonant frequency this gives us the following:

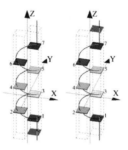

The picture on the left shows us the gate ғ′c′ on the lower part of the system, and the various activated layers within the visual spectrum. Also note the spiral movement as every sequential layer gets activated. The activated layers **1, 3, 5 and 7** lie in the **right front** area, and the activated layers **2, 4 and 6** lie in the **left back** area.

The picture on the right shows us the gate ғ′ғ′ on the upper part of the system, and the various activated layers within the visual spectrum. Here, the activated layers **1, 3, 5 and 7** lie in the **right front** area and the activated layers **2, 4 and 6** lie in the **left back** area.

This leaves us with two more possibilities with potentially different results:

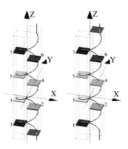

The picture on the left shows us the gate F'C' on the lower part of the system, and the various activated layers within the visual spectrum. Also note the spiral movement as every sequential layer gets activated. The activated layers **1, 3, 5 and 7** lie in the **left back** area, and the activated layers **2, 4 and 6** lie in the **right front** area.

The picture on the right shows us the gate F'F' on the upper part of the system, and the various activated layers within the visual spectrum. Here, the activated layers **1, 3, 5 and 7** lie in the **left back** area and the activated layers **2, 4 and 6** lie in the **right front** area.

These activated areas lie on the opposite sides from the first setting.

To Summarize:

The activated areas are the following:

For layers 1, 3, 5 and 7, activation happens right front, and for layers 2, 4 and 6 it happens left back.

For layers 1, 3, 5 and 7, activation happens left back, and for layers 2, 4 and 6 it happens right front.

We have discussed the eight different areas – four parts of the body, each up and down – and we have concluded that the gate settings can lead to various results. Depending on the chakra, and on the resocle eight gateways to activate a resonant frequency on a chakra.

As an example, we will look at which subsequent energetic

routes need to be used in order to activate subsequent layers. We will look at resonant frequency 1 on the seven chakras and we will investigate resonant frequency 2 on chakra 1. This will give us an idea of the subsequent layers as they get activated, visualized in the spectrum A.

Resonant frequency 1 on chakra 1 activates the following layers:
  layers 1.1.1(red)–1.1.2(orange)–1.1.3(yellow)–1.1.4–1.1.5–1.1.6–
  1.1.7

Resonant frequency 1 on chakra 2:
  layers 2.1.1–2.1.2–2.1.3–2.1.4–2.1.5–2.1.6–2.1.7

Resonant frequency 1 on chakra 3:
  layers 3.1.1–3.1.2–3.1.3–3.1.4–3.1.5–3.1.6–3.1.7

Resonant frequency 1 on chakra 4:
  layers 4.1.1–4.1.2–4.1.3–4.1.4–4.1.5–4.1.6–4.1.7

Resonant frequency 1 on chakra 5:
  layers 5.1.1–5.1.2–5.1.3–5.1.4–5.1.5–5.1.6–5.1.7

Resonant frequency 1 on chakra 6:
  layers 6.1.1–6.1.2–6.1.3–6.1.4–6.1.5–6.1.6–6.1.7

Resonant frequency 1 on chakra 7:
  layers 7.1.1–7.1.2–7.1.3–7.1.4–7.1.5–7.1.6–7.1.7

Resonant frequency 2 on chakra 1:
  layers 1.2.1–1.2.2–1.2.3–1.2.4–1.2.5–1.2.6–1.2.7

Resonant frequency 1 on chakra 1 gives us the following:

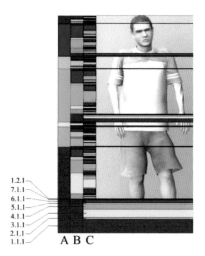

We start with the activated layers via the fixed setting $(c+F+1)$ left back, picture a (below), followed by the activated layers via the fixed setting $(c+F+1)$ left front, picture b, then activated layers via the fixed setting $(c+F'+1)$ right front, picture c. In picture d we see the activated layers via the fixed setting $(c+F'+1')$ right back. From left to right on the pictures we complete the total composition of a+b+c+d, which gives us the layout in the individual energy field spectrum C. In B we have the completion in the human energy field spectrum and in A in the visual spectrum.

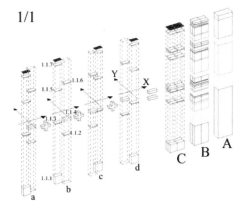

There are four different gateways on chakra 1 that can activate resonant frequency 1. In the picture above, a, b, c and d indicate those four energy lines that can be activated.

- The Gate on the Left Back activates layers 1–3–5–7 on the Left Back and layers 2–4–6 on the Right Front. (a)
- The Gate on the Left Front activates layers 1–3–5–7 on the Left Front and layers 2–4–6 on the Right Back. (b)
- The Gate on the Right Front activates layers 1–3–5–7 on the Right Front and layers 2–4–6 on the Left Back. (c)
- The Gate on the Right Back activates layers 1–3–5–7 on the Right Back and layers 2–4–6 on the Left Front. (d)

In the next picture we visualize, from left to right, the various layers activated by *Resonant Frequency 1 on Chakra 2*. First, there are the layers activated via the gateway right front, picture a, followed by layers activated via the gateway right back, picture b, then layers activated via the gateway left back, picture c, finishing with layers activated via the gateway left front, picture d. In c (individual field), B (human field) and A (visible light spectrum), we get the overall picture within the entire spectrum.

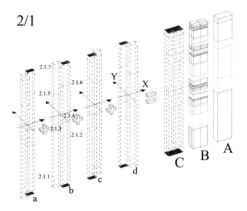

- The Gate on the Right Front activates 1, 3, 5, 7 on the Left Back and 2, 4, 6 on the Right Front.

- The Gate on the Right Back activates 1, 3, 5, 7 on the Left Front and 2, 4, 6 on the Right Back.

- The Gate on the Left Back activates 1, 3, 5, 7 on the Right Front and 2, 4, 6 on the Left Back.

- The Gate on the Left Front activates 1, 3, 5, 7 on the Right Back and 2, 4, 6 on the Left Front.

We now do the same for resonant frequency 1 on chakras 3, 4, 5, 6 and 7.

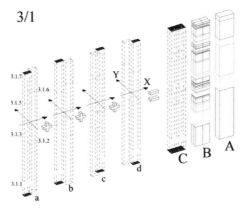

- The Gate on the Left Back activates 1, 3, 5, 7 on the Left Back and 2, 4,6 on the Right Front.

- The Gate on the Left Front activates 1, 3, 5, 7 on the Left Front and 2, 4, 6 on the Right Back.

- The Gate on the Right Front activates 1, 3, 5, 7 on the Right Front and 2, 4, 6 on the Left Back.

- The Gate on the Right Back activates 1, 3, 5, 7 on the Right Back and 2, 4, 6 on the Left Front.

4/1

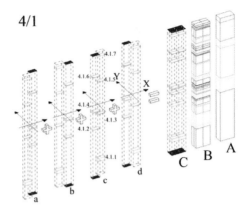

- The Gate on the Right Front activates 1, 3, 5, 7 on the Left Back and 2, 4, 6 on the Right Front.

- The Gate on the Right Back activates 1, 3, 5, 7 on the Left Front and 2, 4, 6 on the Right Back.

- The Gate on the Left Back activates 1, 3, 5, 7 on the Right Front and 2, 4, 6 on the Left Back.

- The Gate on the Left Front activates 1, 3, 5, 7 on the Right Back and 2, 4, 6 on the Left Front.

5/1

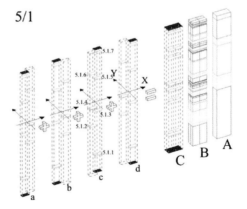

- The Gate on the Left Back activates 1, 3, 5, 7 on the Left Back and 2, 4, 6 on the Right Front.

- The Gate on the Left Front activates 1, 3, 5, 7 on the Left Front and 2, 4, 6 on the Right Back.

- The Gate on the Right Front activates 1, 3, 5, 7 on the Right Front and 2, 4, 6 on the Left Back.

- The Gate on the Right Back activates 1, 3, 5, 7 on the Right Back and 2, 4, 6 on the Left Front.

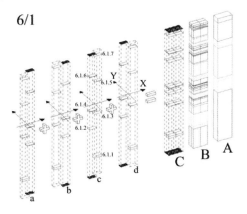

- The Gate on the Right Front activates 1, 3, 5, 7 on the Left Back and 2, 4, 6 on the Right Front.

- The Gate on the Right Back activates 1, 3, 5, 7 on the Left Front and 2, 4, 6 on the Right Back.

- The Gate on the Left Back activates 1, 3, 5, 7 on the Right Front and 2, 4, 6 on the Left Back.

- The Gate on the Left Front activates 1, 3, 5, 7 on the Right Back and 2, 4, 6 on the Left Front.

7/1

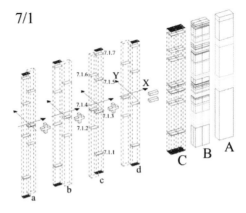

- The Gate on the Left Back activates 1, 3, 5, 7 on the Left Back and 2, 4, 6 on the Right Front.

- The Gate on the Left Front activates 1, 3, 5, 7 on the Left Front and 2, 4, 6 on the Right Back.

- The Gate on the Right Front activates 1, 3, 5, 7 on the Right Front and 2, 4, 6 on the Left Back.

- The Gate on the Right Back activates 1, 3, 5, 7 on the Right Back and 2, 4, 6 on the Left Front.

All that is left to do is to look at the result of putting resonant frequency 2 on chakra 1:

1/2

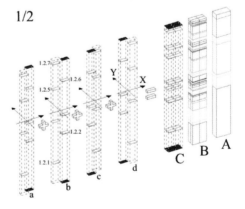

- The Gate on the Right Front activates 1, 3, 5, 7 on the Left Back and 2, 4, 6 on the Right Front.

- The Gate on the Right Back activates 1, 3, 5, 7 on the Left Front and 2, 4, 6 on the Right Back.

- The Gate on the Left Back activates 1, 3, 5, 7 on the Right Front and 2, 4, 6 on the Left Back.

- The Gate on the Left Front activates 1, 3, 5, 7 on the Right Back and 2, 4, 6 on the Left Front.

Next we look at the layers that have been activated by resonant frequency 1 when put on chakras 1, 2, 3 and 4. We use picture a from the previous drawings in our next picture, which shows us in c the place of the various activated layers, activated by complementary gateways. These are: 1.1.1, 2.1.1, 3.1.1 and 4.1.1. The same pattern returns in the orange, yellow, green, blue, indigo and violet parts of the spectrum. The activated areas shift from a left back to a right front one!

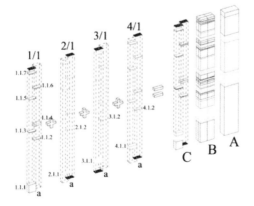

We can complete the picture by visualising the resonant frequency 1 on chakras 5, 6, and 7, and resonant frequency 2 on chakra 1. The trend continues as the subsequent layers, 5.1.1, 6.1.1, 7.1.1 and 1.2.1, switch from left back to right front, even when we move on from resonant frequency 1 on chakra 7 to the next frequency up, which is resonant frequency 2 on chakra 1.

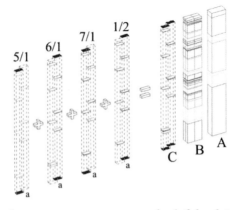

The complementary gateways are: the left back is complementary to the right front, and the right back is complementary to the left front.

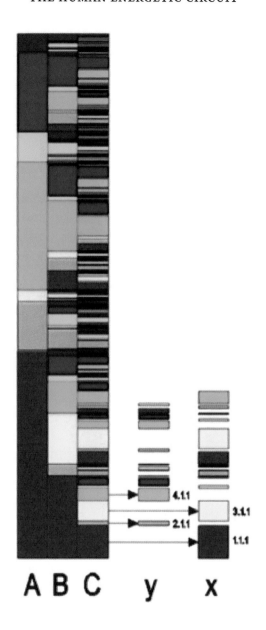

A B C y x

In this way we can visualise the 196 different energy lines, whereby we can connect every part of the body to its energy line.

To finish, I would like to check on the balance between the complementary gateways. Take the drawing of the division of the visible spectrum. Suppose c is the right front of the body. We know that c is a representation of the sites where the chakras manifest. x are the areas that can be connected to a gate on the right front. y are the areas that can be connected to a gate on the left back.

This next picture shows that 1.1.1 (red lower) gets information via the gate on the right front of chakra 1. In contrast, 2.1.1 gets information via the gate on the left back of chakra 2. Furthermore, 3.1.1 gets information via the gate on the right front of chakra 3. And back too: 4.1.1 gets information via the gate on the left back of chakra 4. And so it goes on. This leads to the conclusion that, sequentially, the information is passed through a gate right front and one left back.

x then delivers the following series of chakras via a gate on the right front:

1–3–5–7–2–4–6–1–3–5–7–2–4–6–1–3–5–7

y then delivers the following series of chakras via a gate on the left back:

2–4–6–1–3–5–7–2–4–6–1–3–5–7–2–4–6–1

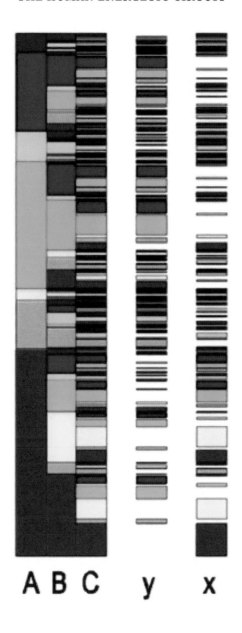

A B C y x

The x and y representation in this picture is a visualisation of these series. x is the series of chakras where the information via a gate on the right front manifests on the right front of the body. y is the series of chakras where the information via a gate on the left back manifests on the right front of the body.

The following picture shows us the balance between the information via a gate on the right front of the chakra (x), and the information via a gate on the left back of the chakra (y), all manifesting on the right front of the body. It is perfectly balanced! But what did you expect?

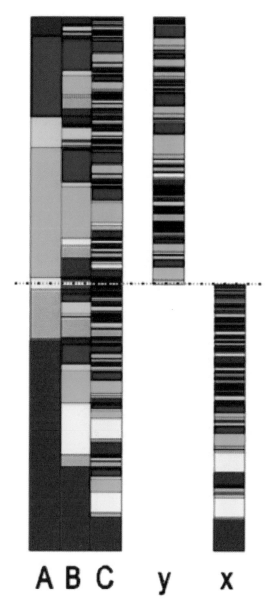

A B C y x

(the picture has a deviation of 0.1%)

Isn't that amazing!

Epilogue
Patrick

Many years ago, when I first studied alternative medical systems, I was struck by two different worlds that appeared to be lightyears apart. On the one hand, there was the physical and the realm of tissues, and on the other hand there was the non-physical, the energetic, the realm of auras, chakras and meridians. Both had explanations for the expression of illnesses, how it happens and why. The trouble, to me, was that the two languages were not compatible. Religion and science, separated through surgical intervention; and the physical and the energetic, separated by ignorance. I kept asking the question: "How can energy be turned into physical tissue, and how does the function of the energy system relate to the function of the tissues?" Nobody could give me a satisfactory answer, not then and not since. However, I somehow had the feeling that it was a pertinent question that needed answering, desperately.

Widening my view on life and the way it functions, I began to get the impression that I would not find the answer in a book or coming out of someone's mouth. It was something I probably had to find out for myself. But how? How would I have the chance to find an answer that had obviously eluded the most brilliant of minds? After breaking out into a sweat of anxiety, I put the thought away, in the sense that I felt that I was going to need a massive amount of help if something like that was ever going to happen.

With no access to the intelligentsia, or the kind of information I would probably need, I left it for what it was. A stupid thought.

Although the thought wasn't in my consciousness daily, it hovered around in the background. So much so that on occasions I realised I was busying myself with just that process. And finally, when the universe is ready, it all starts to unfold. It comes together in a way you couldn't dream up yourself. You couldn't write the kind of scenario that begins to be played out. Ultimately, it isn't about the person who brings out the information; it is about the information itself. As a person, you are only a vehicle, dispensable and finite; the information is what is invaluable and eternal.

So what have I now learned from this journey? In my world – the world I dreamed of being disease-free – I needed to have a better understanding of the process of disease itself. It feels wonderful now to be able to put it into very simple terms:

1. All matter is made directly from energy and displays the major characteristics of the energy field that matter comes from.
2. All diseases are an expression, both energetically and physically, of an imbalance in the way the system is functioning after it has been created, within the limits of its creation.
3. The system can only be pulled out of harmony in two possible ways: pushed inwardly (yin) or pushed outwardly (yang). There are only two types of disease.
4. Each type of disease only has two possible causes. For contracting diseases (yin), either the outer force – the force of the environment – is larger than ususal, or the inner force – the strength of the individual – is too weak. For expanding diseases (yang), either the outer force is too weak or the inner force is too strong. And that is it!
5. When a force is identified as being too strong, we need to weaken it. This means that we need to project ourselves less powerfully, or when it is the force of the outer world, we need to get away from it.

6. When a force is identified as too weak, we need to strengthen
   it. When it is our own inner force, we need to work hard at
   becoming stronger, and when the outer world is too weak, our
   environment could help us by "kicking ass".

We have also learned how energy is transformed from non-physical
to the physical, meaning the tissues, which then expand into organs
and organisms. We learnt to follow the energy lines down into the
physical realm and we saw what type of information is distorted
and where exactly in the tissues it shows up. From this basic information, people can now study life in a totally different way, and
truly understand the natural anatomy and physiology of human
beings and of the universe itself. Rewriting our understanding of
how we are made and how we function will become an essential
part of us growing up.

Humanity is going through an important phase in its evolution.
The most critical part in this is an expansion of our understanding
of what life really is and how it functions. A growing consciousness
will allow human beings to maintain a balanced life in a much
simpler fashion. Things we debate seriously and have wide-ranging
differences of opinions about can now become clear and univer-
sally known. This creates space and time for us to focus on the next
level of schooling we are entering. The world will have changed
forever.

Isn't that great!

Also, the question *"Why Me?"* now has a simple answer:

Because I am made for it. I did not get a say in it; it simply is
the way it is.

And in the same way, we do not have to be afraid of what
appear to be destructive changes. This clears the way for the new
consciousness that you, no doubt, want to be a part of.

I wish you luck and happiness.

www.activehealthcare.co.uk

www. pqliar.net

Epilogue
Erik

After having done more than 15 years of investigation into one of the most mysterious constructions of all time, I feel obliged to spend a few moments reminiscing. One of the questions I have is: "What if you knew then what you know now?" The truth is that I would do it again, in spite of all the pain, physically and mentally, and in spite of all the hardship, in spite of all the arguing. There was simply no other way.

We can compare the investigation to solving a jigsaw puzzle. On the cover of the box there is a nice picture. Inside the box there are a thousand pieces and you display them on the table so that you can have an overview of the kind of pieces you need to work with. Then you start to link the first few pieces together. Gradually, some parts of the picture on the box begin to appear in your creation. You reach a point where it appears that none of the pieces fit onto any of the others anymore, and you are ready to throw them all back into the box. But then someone enters the room and tells you that you haven't turned all the pieces around yet. You get to see more small pieces of the puzzle and you are encouraged by that. When only a few pieces remain on the table, you don't find it hard anymore to finish the puzzle completely and to have the full picture staring at you.

My investigation followed a similar course. When I tried to

decipher what others had learned from the known facts of the pyramid, and I still wasn't able to satisfy my hunger, I knew where to look; the picture on the box, engraved in my mind. I only needed to find the right pieces to place them in their appropriate spots. I had the picture, I just did not know that the road leading to it would be strewn with contradictions and disappointments. Looking back, it is obvious there were so many contradictions, as everything is made out of opposites. Everything encompasses the complementary yin and yang. Every colour holds all the other colours within itself, and has yin and yang entangled in it too. This is life, and the experiences in the school of life are filled with contradictions. Without the contradictions there can be no life. How can you possibly know what "cold" means if you have no notion of what "warm" is? How can you experience "happiness" if you have never been "unhappy"? What is large without small? Thick without thin? What is healthy if you don't know illness?

Meeting Patrick, and later working together, meant a lot to my investigation. To me (as it probably was to him) it meant coming in contact with the opposite approach. I busied myself connecting numbered points via straight lines, as Patrick described it, whilst he was trying to connect disease to energy patterns. These were two completely different worlds and yet we still connected. There was probably an intuitive sensing that there needed to be another side to the coin.

In a way, you can view the entire evolution as solving a jigsaw puzzle. To begin with everything is slow and few combinations are possible. However, the more pieces that are joined together, the quicker it continues and the more results become visible.

Many people, and especially those who are stuck in desperate situations such as terminal diseases, come up with the question: "Why me? What did I do wrong? Others don't try as hard as I do and they don't have to suffer the problems I have."

Me too! I asked myself the question "Why me?" Eventually I learned to accept it for what it is. I tried to see the positive side of it and I began to consider it a privilege to be put on this path. To receive the most beautiful, but also most difficult, of assignments is for a designer the ultimate prize. For myself, I described the assignment as *"To Create Heaven on Earth"*, and when you accept it then there is no other choice but to finish it. What I didn't know was that in order to reach heaven you will also have to experience hell.

Because I moved up to the next consciousness level, and because I have certainly felt it, I realised that there must be a truth behind the whole scheme. I believe very strongly in this consciousness therapy and it saddens me that its use is not more widely spread. As I made the choice, encouraged by William, I started the investigation.

Even before that time I was consciously trying to improve my life and to take responsibility for it. This became even more obvious during the investigation into the visible light spectrum. All I could find was a division into six colours: red, orange, yellow, green, blue and violet. Almost nowhere was indigo mentioned. Including indigo into the equation made a huge difference to the study and to me personally. It seemed as if we had rediscovered something we had lost for some time. We had lost our intuition and our mind. And we need that consciousness in order to let the knowledge grow.

Being born has given us the privilege to experience life at a much slower pace so that we can evaluate and learn. In the meantime, society is driving us forward without letting us pause; keeping us busy without allowing time to look around. This results in us not having time or inclination to ask the question "Why me?". But when we do, we have no other choice but to start the journey towards finding the answer. It is this journey that helps us to become a little "lighter".

The other aspect of this journey is that it teaches you to make your own choices in life because the journey is an individual one. You need to weigh up the pros and cons without actually knowing what the outcome will be. You are travelling. Imagine you are standing at a train station waiting for the train to take you to your destination. Next to you are two suitcases containing everything you need for the journey. After a while, a train approaches the station and people start to become impatient. The train stops. The doors open and it turns out that it is already packed. Everybody is clambering to get on with all of their possessions, but it is impossible. The conductor suggests that everybody can board the train provided they all leave some of their luggage behind. Many people refuse and get off the train to stay with their luggage. They hope to have better luck on the next train that will come along. It is, of course, their choice. But what if there won't be a "next" train to your specific destination?

I boarded that train over 15 years ago. A journey of hit and miss eventually became bearable because you are never alone. Somebody or something in the environment is always there when you need it.

The journey involves a lot of giving in order to be able to receive later. To others it will be a journey of receiving so that they are able to give later. What can be more rewarding than reaching your destination together with others to enjoy a well-earned rest? It is a journey, not a holiday. Doing nothing, being lazy, is not an option here. Once you have reached your destination, you rest a little while, but shortly you will notice that you are already preparing for the next journey. It is all to do with travelling through life itself. In life there are masters and students. A student is not a master yet, but a master will always remain a student.

During the journey you very quickly realise that there is no such thing as a return ticket. There is only a one-way ticket; to keep moving forward. There is no return train and it will never ever run. It will always move forward until you reach your destination. You do not have any other option. The only other option is to jump off of the train, to get out of life.

Sitting on the hard wooden benches of the train, I see my

destination moving closer. But you can never know what is still in store during that last part of the trip. Once I reach the destination I know what I need and want to do. To me, it has become clear; I want to become a conductor on a train. I want to help people and guide them on this type of journey. First we travel in known territory, but gradually we will move further afield. There will always be people waiting at the stations, and maybe one day you will be packed and ready, and on your ticket it simply says:

"Why Me?"

FOLLOW THE PATH AND MAYBE WE WILL MEET IN PARADISE.

Part II of *Why Me?* gives more scientific proof to underwrite everything we have offered and discussed here. Linking everything to ancient wisdom and mythology shows us the truth in the old stories and the truth in the story of Part I.

Truth Stands the Test of Time, and when it talks about the universe, about life itself, it has to be that whatever was true yesterday is still true today and still will be tomorrow.

Stala cola, control,
let us peacefully create the world we
want to live in by living the way
we want to live

over create

Printed in Great Britain
by Amazon

79393084R00181